# SpectrumGuide™

## *Radio Frequency Allocations in the United States, 30 MHz–300 GHz*

*Third Edition*

*Bennett Z. Kobb*

New Signals Press

SpectrumGuide:
Radio Frequency Allocations
in the United States,
30 MHz–300 GHz
Third Edition



ISBN 0-9641546-1-7



New Signals Press Inc.
P.O. Box 435
Falls Church, Virginia 22040–0435
spectrum@newsignals.com

Cover by Bill Bricker

Printed in the USA

Look for SpectrumGuide on the World Wide Web.
E-mail us or check the search services for the current URL.

## Contents

# PREFACE

One hundred and fifty years ago, the best-selling books in America were emigrant's guides describing trails and their landmarks, watering places, and potential dangers to the thousands of people planning or making their ways across our continent.

SpectrumGuide is today's essential equivalent for all of us involved in spectrum-dependent technologies and services. Momentous changes are occurring in spectrum management, and any assumption about the future is fragile. Even so, unimaginable fortunes are being risked (and some are being made) through creative spectrum acquisition and habitation.

Some 1800s emigrant's guides were assembled by people who never reached St. Louis, but this guide has unassailable credibility because, when it comes to making or describing a trail through the FCC Rules, Bennett Kobb has "been there" and "done that."

Today, Congress, the Administration, and a thoroughly revamped and energetic FCC share an agenda, the National Information Infrastructure (NII). The NII will be, in part, spectrum-dependent and will affect spectrum management. In addition to this powerful influence, three other momentous changes are occurring in the spectrum world.

First, new entities are involved: ham operators finding favorite bands suddenly at risk; license winners seeking to understand their new properties; entrepreneurs employing wholly new technologies not considered when present FCC Rules were made; and not least, computer makers realizing that wireless communications provide a powerful feature-set for differentiating products, enlarging markets, and benefiting users.

Second, market forces, as displayed in license auctions, have now assigned billion-dollar cash values to pieces of spectrum, reflecting spectrum scarcity.

Third, paradoxical as it may seem, even while the shortage increases, much more spectrum is becoming accessible. This "new" spectrum comes from mandated turnover of federal spectrum to civilian use and from technology advances that enable better use of spectrum: modulation schemes, digitization, frequency reuse, data compression, and using higher frequency bands, to name but a few. Altogether, useful spectrum is increasing 6–8% per year, including newly-won unlicensed bands discussed fully in this guide.

To navigate these changes, it's essential to understand the rules. To would-be users of spectrum, making one's way through the FCC process resembles the most antagonistic video-game treasure-hunt ever, with unspeakable threats popping up everywhere. The format of the FCC Rules is in part responsible: for any given frequency band the applicable regulations may be strewn throughout five microscopic-print tomes of the Federal Code. Understanding a Rule may also require combing the record of the proceeding in which it was developed, and, often, reports from both the FCC and the NTIA. Failing to find all pertinent material can be fatal.

SpectrumGuide does this research for us. And, along with straightforward facts and clear pointers to all the regulations for each band, the author develops especially insightful commentaries.

Anyone who hopes to exploit or bring about changes in the FCC Rules needs the most effective and dependable guidebook. If SpectrumGuide didn't exist, each of us would have to create our own. We couldn't possibly do it as well.

*James F. Lovette*

# ACKNOWLEDGMENTS

Jim Lovette, principal scientist for network outreach at Apple Computer, and Corwin Moore, administrator of the Personal Radio Steering Group, have been my patient mentors in all things wireless.

This work would not have been possible without the assistance I have received for many years from the Federal Communications Commission, especially the Office of Engineering and Technology, the Office of Public Affairs and the FCC Library; and the staff of the National Telecommunications and Information Administration in the U.S. Department of Commerce. This work does not necessarily represent the views of any government agency.

Thanks to Dr. Andrew Clegg and Dewayne Hendricks for their continued support.

The chief cheerleaders are my parents, Marthe and Sol Kobb, to whom this book is affectionately dedicated.

## HOW TO USE THIS BOOK

The radio spectrum is that portion of the electromagnetic energy spectrum used by radio waves. "Radio services" are categories of radio use. Setting aside frequency ranges, or bands, in the radio spectrum to radio services is a government function called spectrum allocation or frequency allocation.

The nations of the world, through the International Telecommunication Union (ITU), allocate spectrum to radio services in the form of an International Table of Allocations. Each nation also establishes a domestic table of allocations based on the International Table. International allocations are usually, but not always, incorporated into domestic allocations.

SpectrumGuide contains the 30 megahertz (MHz) to 300 gigahertz (GHz) portion of the United States Table of Frequency Allocations (the "Table") as administered by the Federal Communications Commission (FCC) and the National Telecommunications and Information Administration (NTIA).

The book spans the Very High Frequency (VHF, 30–300 MHz), Ultra High Frequency (UHF, 300–3000 MHz), Super High Frequency (SHF, 3–30 GHz) and Extremely High Frequency (EHF, 30–300 GHz) bands. Most commercial activity is in these bands.

To this official data, we add descriptive narratives about spectrum uses and trends. *SpectrumGuide emphasizes bands allocated to services instead of the individual frequencies assigned to millions of stations.*

Congress chartered the FCC, an independent agency, to regulate non-federal use of radio. FCC spectrum allocations usually involve a public comment process.

The federal government's use of radio is subject to Presidential authority. This authority is exercised through the NTIA and its Administrator, the Assistant Secretary of Commerce for Communications and Information.

The FCC and NTIA maintain separate versions of the Table to serve their particular requirements. One agency may perceive a spectrum range to be a single band, for example, while the other perceives it to contain several bands. One may have changed its table while the other has yet to reconcile its table with the other's changes.

There is unlikely to be a single, uniform national table until the two agencies publish the same data in the same format and update it at the same time. We used both agencies' material to compile SpectrumGuide.

In U.S. radio allocations, the terms "government" and "non-government" ordinarily refer to the use (or non-use) of bands by the U.S. federal government. State and local governments are not considered "government" users in the Table of Allocations. The FCC provides, however, special frequency assignments for state and local government use, particularly in the Public Safety Radio Services.

The U.S. Table consists of frequency bands, lists of the government and non-government services permitted to use the bands, and footnotes that modify the rights of a service to operate in a band.

Footnotes may, for example, add allocations, may add protection or remove protection from interference for a particular service or radio user.

Footnotes are designated "G" for government allocations and "NG" for non-government allocations. "US" footnotes apply to both government and non-government allocations. Footnote numbers not accompanied by letters indicate international footnotes that are included in the U.S. Table. International footnotes occasionally refer to other documents in the International Radio Regulations.

Footnotes are sometimes associated with groups of bands even though the footnotes actually apply only to specific frequencies within a band or to a subset of bands within the group. Read a footnote carefully to determine whether or not it actually applies to the band of interest.

SpectrumGuide cannot be considered an official government document. Readers should rely on the official regulations of the FCC, NTIA and ITU when planning, constructing or operating communications systems.

All material in SpectrumGuide is based on public sources, including declassified defense sources. Some U.S. government spectrum uses and users remain classified.

A station that operates in a "primary" service has full rights to operate in bands allocated to it, under rules that apply to the service and the terms of its license.

By international convention, primary services appear in CAPITAL letters. For example, "FIXED" means that the fixed service (radio communications between stationary stations) is primary in a given frequency band. The FCC further resolves general descriptions such as "FIXED" into individual radio services. For example, one of the Fixed Services is the Private Operational Fixed Service used by commercial, state and local government licensees.

A station in a primary service is legally protected against interference from a station in a "secondary" service. A station in a secondary service cannot claim interference protection from stations in a primary service. Secondary services are portrayed in upper and lower case letters.

Radio services are often primary in some bands and secondary in other bands. A "shared" allocation is one available to two or more services, typically in a mixture of primary and secondary use. When two or more services share the same bands, they often must coordinate their use to minimize interference.

Here is part of a typical band listing:

---

## *10.7-11.7 GHz*

Gov: US211

Non-Gov: FIXED. FIXED-SATELLITE (space-to-Earth). 792A US211 NG41 NG104

FCC: DOMESTIC PUBLIC FIXED (21). PRIVATE OPERATIONAL FIXED MICROWAVE (94).

In this example, "Gov: US211" means that footnote US211 states requirements that apply to federal government (Gov) use of the 10.7–11.7 GHz band.

"Non-Gov: FIXED. FIXED-SATELLITE (space-to-Earth). 792A US211 NG41 NG104" means that non-government use of this band is for fixed service stations and fixed satellite transmissions to Earth. Capital letters indicate that these services are primary in this band. The footnotes list international, U.S. and non-government requirements that apply to this band.

The line "FCC: DOMESTIC PUBLIC FIXED (21). PRIVATE OPERATIONAL FIXED MICROWAVE (94)." is called an "FCC Use Designator". It indicates the non-government services authorized to use the band, with the FCC Rule Part number. FCC Rules appear in the Code of Federal Regulations and the Federal Register. Not all bands include this designator.

Other than in critical features such as the designation of primary and secondary services by letter case, the government is not consistent in style, punctuation or the order of items within entries in the Table of Allocations.

Many radio services use multiple bands. Sometimes these bands can be separated by many megahertz. Land mobile radio systems, for example, may use separate bands for the base station and the mobile radios.

You can find the principal narrative for a service in the listing for the lowest frequency band that it occupies.

In SpectrumGuide, a range of allocated frequencies is a "band". Two or more bands are "spectrum" and frequency ranges within a band are "segments".

The United States participates in World Radio Conferences which modify the International Table. Occasionally the FCC and NTIA update their allocations and footnotes based on international developments. We endeavor to present information that conforms with current U.S. regulations at the time of publication.

As always, you're invited to contribute feedback and updated information for the next edition.

# Frequency Allocations

## 30-30.56 MHz

Gov: FIXED. MOBILE.

Non-Gov: None

This is one of several "low-band" VHF federal fixed and mobile bands. These are less used than higher frequency bands because of interference caused by long-distance propagation of signals, especially during high sunspot activity. Higher bands are not so affected.

The largest user in this band is the Army for mobile dispatch operations, such as training, security, maintenance and administration. Other military services and federal agencies make similar use of these low VHF bands, especially the Air Force, Navy and the Interior and Agriculture Departments.

## 30.56-32 MHz

Gov: None

Non-Gov: LAND MOBILE. NG124

FCC: PRIVATE LAND MOBILE (90).

This and other low-band VHF bands are available to several Private Mobile Radio Services (PMRS), also called Private Land Mobile Radio (PLMR) services. These are categories of not-for-profit one- and two-way radio for licensees' internal use. They typically are used for voice and computer-aided dispatch of vehicles and for paging.

The FCC is changing the terminology it traditionally has used to classify radio services. The new terminology is somewhat more opaque. An essential distinction is between PMRS and the Commercial Mobile Radio Services (CMRS). Licensees of CMRS businesses provide for-profit, telephone-interconnected (mobile and portable telephone or dial-up paging) service to the public.

Examples of CMRS include private paging services (except for not-for-profit, internal paging systems); Business Radio Service systems, Specialized Mobile Radio services and 220–222 MHz land mobile systems, to the extent all of these offer for-profit, telephone-interconnected service to the public; Public Mobile Services such as the Paging and Radiotelephone Service and the Cellular Radiotelephone Service; the 454 MHz and 800 MHz air-ground radiotelephone services; the Offshore Radiotelephone Service; most types of mobile satellite services; and Personal Communications Services.

Previously, most for-profit mobile services were called "radio common carriers". This term has not completely disappeared from use.

Recent legislation brought all for-profit mobile radio services under the CMRS umbrella. Some former PMRS services were reclassified as CMRS because they are largely devoted to for-profit service with interconnection to the public telephone network. In some cases PMRS and CMRS operations share bands.

Recent deregulation by the FCC permits CMRS operators to provide fixed service on a co-primary basis. This means that cellular and PCS carriers, for example, could offer "wireless local loop" fixed telephone service in competition with local telephone companies.

This wireless local loop service may be the means by which meaningful local exchange competition will emerge.

The FCC grants PMRS licenses only to individuals, organizations or agencies that qualify under eligibility rules developed over decades of proceedings. Trade associations representing each service category perform quasi-licensing functions and frequency coordination under FCC oversight. The PMRS encompass a diverse array of services.

The Business Radio Service is the largest private mobile radio service, with broad eligibility for business, health care, educational and other institutional users.

The Public Safety Radio Services include the Police, Fire, Highway Maintenance, Forestry-Conservation, Local Government and Emergency Medical Radio Services.

Road construction and repair crews use the Highway Maintenance Radio Service. The Forestry-Conservation Radio Service is used in game law enforcement, forest fire protection, reforestation, flood and erosion control.

The Local Government Radio Service is for state, county and municipal agencies.

The Emergency Medical Radio Service is for the use of physicians, medical technicians and rescuers directly involved in life support, for example in transporting injured persons to care facilities.

The Special Emergency Radio Service is not officially one of the Public Safety Radio Services. It governs other types of assistive radio, such as veterinary radio, radio use by the disabled, disaster relief and school bus dispatch.

The Radiolocation Radio Service provides for all non-federal use of radar.

The Industrial Radio Services include the Power Radio Service for electric, gas and water utilities; the Petroleum Radio Service for oil and natural gas industries; the Forest Products Radio Service for the lumber and paper industries; the Film and Video Production Radio Service for movie and TV producers; the Relay Press Radio Service for newspapers; the Special Industrial Radio Service for construction, farming and mining; the Manufacturers Radio Service for communications at factories; and the Telephone Maintenance Radio Service for telephone repair crews.

The Land Transportation Radio Services include the Motor Carrier Radio Service for bus and trucking companies; the Taxicab Radio Service; the Railroad Radio Service and the Automobile Emergency Radio Service for tow trucks.

The FCC will consolidate the long list of PMRS services into a shorter, simpler list. This consolidation is part of its "refarming" program to make more efficient use of the traditional PMRS bands below 512 MHz.

Refarming is necessary because of the extremely uneven use of the PMRS bands. Channels in some services are congested while channels in other services are lightly used. Each PMRS radio channel accommodates hundreds of licensees and thousands of transmitters.

With the explosive growth in business and public safety mobile communications, demand for channels is much greater than the supply.

"We are convinced that, without significant regulatory changes in the bands below 512 MHz, the quality of PLMR communications will likely deteriorate to the point of endangering public safety and the national economy," the FCC said.

Refarming entails radical technical and regulatory changes. New spectrum efficiency standards will be imposed over time, reducing the power and bandwidth of private mobile radio transmitters.

Current PMRS channels are generally spaced as follows:

| *Spectrum (MHz)* | *Spacing (kHz)* |
|---|---|
| 25–50 | 20 |
| 72–76 | 20 |
| 150–174 | 15 |
| 220–222 | 5 |
| 421–430 | 25 |
| 450–470 | 25 |
| 470–512 | 25 |
| 806–821 / 851–866 | 25 |
| 821–824 / 866–869 | 12.5 |
| 896–901 / 935–940 | 12.5 |

Under its refarming rules, the FCC will list channels every 7.5 kHz in the 150–174 MHz VHF band and every 6.25 kHz in the 421–430, 450–470 and 470–512 MHz UHF bands. The agency will authorize new radios for market that are capable of operating on the narrower bandwidths, or that provide equivalent efficiency through such means as Time-Division Multiple Access.

The FCC's goals for refarming include a fourfold increase in the number of channels available in PMRS, as well as some means of obtaining revenues for use of the spectrum in the form of auctions or spectrum use fees. Licensees already pay application fees, regulatory fees, frequency-coordination fees and service fees in addition to equipment costs, however.

Refarming should promote entrepreneurship. Some licensees could purchase exclusive use of their channels at auction. The exclusive licensees could then sell communications service to customers.

There are actually two types of exclusivity. In some services, exclusivity means that there are no other co-channel licensees within a certain radius of the base station.

In the PMRS bands below 512 MHz, however, exclusivity refers to the ability of the exclusive licensee to prevent any additional licensees from using the spectrum. Existing licensees would be permitted to remain on their channel on a co-primary basis with the exclusive licensee.

The refarming transition may be lengthy and difficult for the private mobile radio industry. Some commenters believe that even with refarming, additional spectrum allocations to private mobile radio will be needed.

Several trade associations have joined a “Coalition of Private Users of Emerging Multimedia Technologies” (COPE) to advocate allocations for new uses of mobile radio such as image and high-speed data communications.

“An analysis based on an estimate of PLMR transmitters in a major urban area (e.g., New York or Los Angeles) confirms that refarming all PLMR spectrum will at best provide sufficient capacity to support the growth of traditional voice and data dispatch during the 1990s but not new emerging information technologies,” the group said.

COPE told the FCC that at best, the refarming of the PLMR spectrum below 470 MHz will only result in moderate relief in PLMR congestion, and if PLMR continues to grow, these gains would be eviscerated.

The FCC also has launched a large scale examination of the Public Safety Radio Services, looking toward providing increased capacity and capability. The Commission cited several bands that might be reallocated for PSRS use, including 152–152.255, 152.495–152.855, 157.755–158.115, 335.4–399.9 and 2110–2120 MHz.

“Today, most radio communications by public safety agencies are conventional operations involving the transmission of voice and data," the FCC said. "The demands on public safety communication systems, however, are expanding to include a host of high speed data applications such as fingerprints, photographs, building diagrams, slow and full motion video, and decisional data."

Various channels in the low VHF spectrum could be made available for a Local Area Determination Service. This experimental service currently is used to provide golfers with location information on the course; for surveying; mapping of radioactive surface contamination and the precise locating of artifacts in archaeology.

## 32-33 MHz

Gov: FIXED. MOBILE.

Non-Gov: None

This band is for federal agency and Army, Navy, Marine and Air Force fixed and mobile communications.

## 33-34 MHz

Gov: None

Non-Gov: LAND MOBILE. NG124

FCC: PRIVATE LAND MOBILE (90).

The Special Emergency, Highway Maintenance, Business, Petroleum, and the Fire Radio Services use this part of the low VHF land mobile radio band.

## 34-35 MHz

Gov: FIXED. MOBILE.

Non-Gov: None

This band is for federal agency and Army, Navy and Air Force fixed and mobile communications.

## 35-35.19 MHz

Gov: None

Non-Gov: LAND MOBILE. NG124

FCC: PRIVATE LAND MOBILE (90).

This primarily Business Radio Service band also contains frequencies used for telephone company maintenance communications, for small paging systems authorized to handicapped persons, and for developmental stations.

## 35.19-35.69 MHz

Gov: None

Non-Gov: LAND MOBILE. NG124

FCC: DOMESTIC PUBLIC LAND MOBILE (22). PRIVATE LAND MOBILE (90).

This band contains frequencies for the Special Industrial and Special Emergency Radio Services and the low-band VHF base station frequencies for the Paging and Radiotelephone Service, one of the Public Mobile Services.

## 35.69-36 MHz

Gov: None

Non-Gov: LAND MOBILE. NG124

FCC: PRIVATE LAND MOBILE (90).

This band is allocated to the Business and Special Industrial Radio Services and for developmental stations used to explore radio technology.

## 36-37 MHz

Gov: FIXED. MOBILE. US220

Non-Gov: US220

This general fixed and land-mobile band is used by federal agencies and the Army, Navy and Air Force.

The band also contains one of two low-band VHF frequencies used in oil spill cleanup and containment.

## 37-37.5 MHz

Gov: None

Non-Gov: LAND MOBILE. NG124

FCC: PRIVATE LAND MOBILE (90).

This band is mostly devoted to the Police Radio Service. It contains a few frequencies for the Forest Products and Power Radio Services, and developmental stations.

## 37.5-38 MHz

Gov: Radio Astronomy. 547

Non-Gov: LAND MOBILE. Radio Astronomy. 547 NG59 NG124

FCC: PRIVATE LAND MOBILE (90).

This band is allocated to the Power Radio Service (electricity, gas, water distribution), and to the Public Safety Radio Services.

Radio astronomers use 37.5–38.25 MHz to study the Sun, Jupiter, stars and interstellar clouds.

The 37.5–38.25 MHz spectrum is one of many restricted bands in which the FCC's Part 15 rules permit unlicensed devices to emit only very low level emissions.

## 38-38.25 MHz

Gov: FIXED. MOBILE. RADIO ASTRONOMY. 547 US81

Non-Gov: RADIO ASTRONOMY. 547 US81

This is part of the 37.5–38.25 MHz radio astronomy band. Footnote US81 provides that military transportable and mobile stations will continue to use the band.

The 37.5–38.25 MHz spectrum is one of many restricted bands in which the FCC's Part 15 rules permit unlicensed devices to emit only very low level emissions.

## *38.25-39 MHz*

Gov: FIXED. MOBILE.

Non-Gov: None

This band is for federal agency, Army and Air Force fixed and mobile communications.

## *39-40 MHz*

Gov: None

Non-Gov: LAND MOBILE. NG124

FCC: PRIVATE LAND MOBILE (90).

This band is devoted to police and local government communications.

## *40-42 MHz*

Gov: FIXED. MOBILE. 548 US210 US220

Non-Gov: 548 US210 US220

This band is for federal agency, Army, Navy and Air Force fixed and mobile use. The frequency 40.5 MHz is for military distress, rescue, guard and interservice use.

The band contains one of two low-band VHF frequencies for oil spill cleanup and containment. It also is available for wildlife and ocean buoy tracking and telemetering.

The 40.66–40.7 MHz segment is one of several Industrial, Scientific and Medical (ISM) bands. It is available for use by remote controlled devices such as alarms and door openers.

ISM refers to the use of radio energy to perform some function other than telecommunications. Plenty of telecommunications devices use ISM spectrum anyway.

## 42-43.19 MHz

Gov: None

Non-Gov: LAND MOBILE. NG124 NG141

This band is largely devoted to state police radio, with some frequencies for business and industrial use.

## 43.19-43.69 MHz

Gov: None

Non-Gov: LAND MOBILE. NG124

FCC: DOMESTIC PUBLIC LAND MOBILE (22). PRIVATE LAND MOBILE (90).

This band is paired with 35.19–35.69 MHz for private industrial and emergency communications and the Public Mobile Paging and Radiotelephone Service.

## 43.69-46.6 MHz

Gov: None

Non-Gov: LAND MOBILE. NG124 NG141

FCC: PRIVATE LAND MOBILE (90).

The late lamented TV channel 1 occupied 44–50 MHz, but the FCC reallocated that segment to mobile communications in the 1950s.

The 43.69–46.6 MHz band is devoted to land transportation users, police and fire, forestry conservation and state and local governments. The band also includes cordless telephone base station channels 1-15.

Available frequencies for cordless phones shifted in 1984 to ten channels at 46.61–46.97 (base stations) and 49.67–49.97 MHz (handsets). The base stations previously used the noise-plagued 1.625–1.8 MHz band, now part of the AM broadcast band. Handsets used 27 MHz but were shifted to 49 MHz.

Millions of 49 MHz / 1.7 MHz phones were sold. They became infamous for making squawking noises and ringing without being called. We have seen letters to the FCC from irate consumers whose cordless phones rang even before they were taken out of the package.

Industry groups devoted special attention to understanding the weird ability of these phones to spontaneously dial emergency 911 services. The problem eventually was traced to spurious responses of the telephone to random radio noise bursts and signals from other co-channel phones. This resulted in random switchhook closures that the telephone network interpreted as dialing sequences.

In what Commissioner Ervin Duggan called "an act of stunning, even dangerous naivete," the FCC declined for years to require features in cordless phones that would have reduced these problems.

The all-VHF cordless phones include required security features and experience fewer such problems than older models. The multiple channels make trunking (automated channel selection) feasible, more efficiently using the limited spectrum available.

Even so, in high-density areas the original 46 / 49 MHz channels became saturated, leading to blocking (no clear channels available) and crosstalk (audible interference from units nearby). Returns for cordless phones reached nearly three times the average return rate for all consumer electronics, except for home computers and video games.

Aggravating the problem are baby monitors, which the FCC allows to use 49.82–49.90 MHz. They transmit infant sounds to parents elsewhere in the home. Baby monitors transmit any sound within microphone range, and typically are left on continuously. Recreational monitoring of neighbor conversations broadcast by baby monitors is a popular underground pastime.

The perceived shortage of frequencies for cordless telephones has long been of concern to the Telecommunications Industry Association (TIA) which lobbies the FCC on behalf of phone manufacturers.

TIA asked the FCC to add 30 frequencies at 43.72–44.48 (base) and 48.76–49.5 MHz (handsets) to cordless phones. At the low power radiated by these products—roughly 25 microwatts—TIA believed they would not interfere with private land mobile radios operating in these bands.

The FCC granted the request in April 1995. To further prevent interference with land mobile systems, phones using the new frequencies must incorporate mechanisms that prevent use of occupied frequencies.

The new frequencies are not paired; any new base transmitter frequency may be used with any new handset transmitter frequency.

Cordless telephones increasingly are using the 902–928 MHz Industrial, Scientific and Medical (ISM) band. TIA believes that "products operating in that [902–928 MHz] band are clearly differentiated from those at 49 MHz by both price and performance. Initially, products in the 902–928 MHz band will serve primarily business customers and upscale consumers. With time, the costs of ISM products will drop and the market will grow, but the 49 MHz units will continue to provide affordable cordless telephone service for the majority of consumers, including the disabled, as well as elderly people living on fixed incomes."

The 46 / 49 MHz phone is informally known internationally as CT0. CT1 refers to analog UHF cordless phones sold in Europe. CT2 and its successors are European digital cordless telephone standards intended to improve on CT1.

CT2 helped foster the Personal Communications Services (PCS) allocations (see 1850–1990 MHz). Cordless phone systems for business use are now using unlicensed PCS spectrum at 1920–1930 MHz.

This spectrum also supports meteor burst data communications. This technology reflects signals from the trails of millions of micrometeors that shower into the upper atmosphere. The pioneering meteor burst system is Snotel, the U.S. Department of Agriculture's network of precipitation monitoring stations in 12 western states.

## 46.6–47 MHz

Gov: FIXED. MOBILE.

Non-Gov: None

This band includes cordless phone base station channels 16–25 (see 43.69–46.6 MHz).

The FCC and National Telecommunications and Information Administration agreed on continued use of this federal band (and 49.6–50 MHz) for these products.

## 47–49.6 MHz

Gov: None

Non-Gov: LAND MOBILE. NG124

FCC: PRIVATE LAND MOBILE (90).

This band is allocated to public-safety radio, especially for state highway maintenance including self-powered vehicle detectors, and industrial uses in the Forest Products, Special Industrial, Petroleum and Power Radio Services.

The band also includes cordless telephone handset channels 1–15 (see 43.69–46.6 MHz) and the perennial, if obsolete, national Red Cross frequency 47.42 MHz.

## *49.6-50 MHz*

Gov: FIXED. MOBILE.

Non-Gov: None

The band, otherwise allocated to the federal government, is used for cordless phone handset channels 16–25 (see 43.69–46.6 MHz).

FCC Part 15 rules permit other unlicensed devices, including home-built devices, to use 49.82–49.90 MHz. The FCC started to migrate them to 49.82–49.90 MHz in 1976 after contention between then-licensed Citizens Band operations and unlicensed devices in the 27 MHz spectrum.

## *50-54 MHz*

Gov: None

Non-Gov: AMATEUR.

FCC: AMATEUR (97).

The U.S. recognizes the Amateur Radio Service as a voluntary, non-commercial radio service that provides emergency communications; contributes to the advancement of the radio art; expands the national pool of trained radio personnel and fosters international goodwill.

Radio amateurs (hams), obtain FCC licenses through qualifying examinations at several levels of proficiency.

Nicknamed the Magic Band, this six-meter 50–54 MHz band holds a special fascination for radio amateurs. It is devoid of distant signals most of the time. It is famous, however, for sudden signal propagation by sporadic-E, meandering clouds of metallic ions in the E-layer of the ionosphere about 70 miles above the Earth.

These clouds act as mirrors, reflecting even low-power 50 MHz signals across thousands of miles and causing the band to break out in a frenzy of stations.

Propagation by auroral effects also is characteristic of these frequencies. The aurora is associated with solar activity that affects the Earth's magnetic field. Tropospheric bending of 50 MHz signals also occurs, caused by atmospheric changes between air masses of different temperature and humidity. These phenomena can bring distant interferers to VHF TV and mobile radio.

Six meters is home to a subculture of amateur operators who enjoy the band's surprising behavior. Typical uses include single-sideband voice, Morse code, data and FM mobile radio with repeaters. The FCC issued special permission to some amateurs for spread spectrum experiments in this band.

## 54–72 MHz

Gov: None

Non-Gov: BROADCASTING. NG128 NG149

FCC: RADIO BROADCAST (TV) (73). Auxiliary Broadcasting (74).

This band contains TV channels 2, 3 and 4.

The television allocations occupy 72 MHz in the VHF bands. TV stations are expected to eventually vacate these VHF channels and the FCC will reallocate them to other uses. The FCC proposed to relocate all U.S. TV broadcasting to channels 7–51 in the transition to Digital Television (see 470–512 MHz).

# 72-73 MHz

Gov: None

Non-Gov: FIXED. MOBILE. NG3 NG49 NG56

FCC: DOMESTIC PUBLIC LAND MOBILE (22). PERSONAL (95). PRIVATE LAND MOBILE (90).

The bands from 72–73, 74.6–74.8, 75.2–75.4 and 75.4–76 MHz have such diverse uses as emergency fire call boxes, narration of golf tournaments and horse races, local communications at industrial sites, the remote control of model aircraft and Public Mobile transmitters, and radio-aided education of the hearing-impaired.

Common terminology refers to the "72–76 MHz band". This term does not include the 73–74.6 MHz radio astronomy or the 74.8–75.2 MHz aeronautical marker-beacon bands.

Private Mobile Radio Services share 72–76 MHz for both fixed and mobile operations. These include the Fire Radio Service for call boxes, and the Forest Products, Railroad, Special Industrial, and Manufacturers Radio Services for low-power communications only within the confines of a work area such as a rail yard, factory or construction site.

Public Mobile Services stations have channels in 72–73 and 75.4–76 MHz for point-to-point operation, for such uses as control of paging systems.

Auditory assistance devices use 72–73 and 75.4–76 MHz to transmit teachers' speech to students who wear hearing-aid-like receivers, or to loudspeakers distributed throughout a classroom. These systems also are used in theaters, conference centers, churches and other assembly areas.

According to Karen Strauss of the National Association of the Deaf, "the 72 to 76 MHz band, used by FM auditory assistance devices, has been inundated with interference from pagers, cellular phones, emergency dispatch vehicles, electronic equipment, CB/ham radio transmitters, walkie-talkie radios, and vehicle dispatchers. The interference from such high powered users has made these auditory assistance devices virtually unusable in a multitude of settings, the most serious of which is our schools."

Trying to find a better home for auditory devices, the FCC created a Low Power Radio Service at 216–217 MHz but concerns still remain over interference.

The 72.44–75.6 MHz spectrum contains channels used to broadcast narration of sporting events to spectators who rent special portable radios.

The Radio Control (R/C) Radio Service, a Personal Radio Service in Part 95 of the FCC Rules, provides 50 channels in 72–73 MHz for the remote control of model aircraft and 30 channels in 75.4–76 MHz for model boats and cars.

## 73-74.6 MHz

Gov: RADIO ASTRONOMY. US74

Non-Gov: RADIO ASTRONOMY. US74

This exclusive allocation to radio astronomy is used to study radio emissions from the Sun, Jupiter, stars, galaxies and interstellar clouds. It is one of many restricted bands in which the FCC's Part 15 rules permit unlicensed devices to emit only very low level emissions.

## 74.6-74.8 MHz

Gov: FIXED. MOBILE. 572 US273

Non-Gov: FIXED. MOBILE. 572 US273

FCC: PRIVATE LAND MOBILE (90).

See 72–73 MHz for the uses of this band.

## 74.8-75.2 MHz

Gov: AERONAUTICAL RADIONAVIGATION. 572

Non-Gov: AERONAUTICAL RADIONAVIGATION. 572

FCC: AVIATION (87)

This band contains beacons in the Instrument Landing System (see 108–117.975 MHz). It is one of many restricted bands in which the FCC's Part 15 rules permit unlicensed devices to emit only very low level emissions.

## 75.2-75.4 MHz

Gov: FIXED. MOBILE. 572 US273

Non-Gov: FIXED. MOBILE. 572 US273

FCC: PRIVATE LAND MOBILE (90).

See 72–73 MHz for the uses of this band.

## 75.4-76 MHz

Gov: None

Non-Gov: FIXED. MOBILE. NG3 NG49 NG56

FCC: DOMESTIC PUBLIC LAND MOBILE (22). PERSONAL (95). PRIVATE LAND MOBILE (90).

See 72–73 MHz for the uses of this band.

## 76-88 MHz

Gov: 576

Non-Gov: BROADCASTING. 576 NG128 NG129 NG149

FCC: RADIO BROADCAST (TV) (73). Auxiliary Broadcasting (74).

This band contains TV channels 5 and 6.

TV stations are expected to eventually vacate these VHF channels and the FCC will reallocate them to other uses. The FCC proposed to relocate all U.S. TV broadcasting to channels 7–51 in the transition to Digital Television (see 470–512 MHz).

## 88-108 MHz

Gov: US93

Non-Gov: BROADCASTING. US93 NG2 NG128 NG129

FCC: RADIO BROADCAST (FM) (73). Auxiliary Broadcasting (74).

U.S. FM radio broadcasting is allocated 100 channels from 88.1 to 107.9 MHz, including main transmitters and translator stations that extend coverage. The FCC allots groups of FM channels to geographic areas. Non-commercial educational FM stations operate in a reserved subband, 88.1–91.9 MHz.

FM broadcasting is undergoing a digital transformation in the 88–108 MHz band and higher in the spectrum. Digital program content, digital audio production and storage, and digital transmission technologies are lowering costs, improving audio quality and allowing stations to offer entirely new services.

Many FM stations use the Radio Broadcast Data System (RBDS, also called Radio Data System or RDS) to transmit short messages at 1.2 kbps over subcarriers: signals subsidiary to the main FM signal, but within its authorized bandwidth.

With RBDS, a station can display its callsign, program format, and even the name of the current song and artist, on the receiver or other displays such as a highway billboard. Differential Global Positioning System (GPS) corrections can be transmitted via RBDS (see 1215–1240 MHz).

The broadcast industry already is studying data technologies with greater capability than RBDS. The

National Radio Systems Committee of the National Association of Broadcasters is testing systems and will recommend a high-data-rate standard.

The committee is examining systems that have as high as 20 times the data rate of RBDS and that could offer more complex types of information. This technology is launching in Japan, where consumer radio receivers with large display screens are sold.

Beyond ancillary data is the issue of digital improvements in the main broadcast. The industry is moving toward replacing FM with digital modulation and, optionally, integrating it with satellite radio.

This is known in most of the world as Digital Audio Broadcasting (DAB) and in the U.S. as Digital Audio Radio Services (DARS, see 2310–2360 MHz).

The FCC considers DARS to be "an all-inclusive term that encompasses both terrestrial digital audio broadcasting (DAB) and digital audio provided by satellite." A key benefit of DARS will be its ability to overcome the multipath signal fading common in FM.

The Electronic Industries Association is testing several forms of terrestrial DARS from different proponents. A leading candidate is USA Digital Radio, which delivers digital transmission for both AM and FM stations in-band and on-channel (IBOC)—within the station channel already assigned for analog broadcasting. USA Digital Radio was first demonstrated in April 1995.

Other nations believe that IBOC DARS will fail and that America will be left behind the rest of the world. European and Canadian broadcasters advocate Eureka, a DAB system in other spectrum (see 1435–1530 MHz).

Emergency broadcasts also are in a digital revolution. The FCC requires AM, FM and TV stations to adopt a new

Emergency Alert System (EAS). It will replace the Emergency Broadcast System (EBS) in a few years.

EBS was intended to warn the public of enemy attack and natural disasters. It was the successor to the CONELRAD system of the Cold War era and the failed Public Emergency Radio system of the 1970s (see 162.0125–173.2 MHz).

The familiar and irritating EBS Attention Signal (853+960 Hz) will be shortened and accompanied by digital bursts as well as voice announcements in the new EAS. The data includes detailed information about the type and source of emergency.

EAS decoders monitor multiple inputs, such as phone lines and radio receivers, for redundancy. They alert broadcast operators if they detect a genuine emergency message.

The EAS is at the President's disposal for the national Emergency Action Notification (EAN). The highest level of emergency, transmissions of EAN priority have never been made by the White House.

The President may use the EAN to commandeer radio and TV without local intervention, although it is not clear if broadcasters will eagerly implement this capability.

Local and state officials may also originate EAS alerts. Broadcasters have expressed their concerns to the FCC that trigger-happy local officials could interrupt radio and TV with alarmist or political messages.

EAS can address discrete geographic units, including individual homes. It can activate radio and TV sets that are switched off, if the receivers contain special EAS circuitry—and provided that owners program their sets to respond to the many types of EAS alerts.

The FCC's gushing prediction that this circuitry will appear in radio and TV sets has not materialized. Set manufacturers are not required to implement it. EAS user programming is a daunting human interface challenge.

The technology is available in a few NOAA Weather Radio receivers. The Weather Radio VHF stations are installing EAS-compatible warning origination equipment (see 162.0125–173.2 MHz).

In this 88–108 MHz band, Non-Government Footnote NG2 permits "Facsimile Broadcasting Stations". This is a forgotten relic of plans of the 1950s–60s to transmit newspapers by radio. No such stations exist, but the FCC dutifully supplied license application forms for fax stations through the 1980s.

The electronic fax newspaper was a typical element in "home of the future" scenarios. Criswell—the eccentric prophet and columnist portrayed in the Oscar-winning movie "Ed Wood"—predicted: "After 1980, newspapers will be fed into the home by television transmission, where hard copies of print-outs will roll out of the set each morning and afternoon, complete with photos, features, etc. And, because television is under government control, every newspaper will be under government control also."

Criswell also predicted, in 1968, that the capital of the U.S. would move to Wichita, Kansas by 1983; Denver would be turned to jelly by an alien space ray in 1989; anti-gravity technology from Nebraska would power all cars by 1995; and the world will end on Criswell's birthday, August 18, 1999.

---

## 108-117.975 MHz

Gov: AERONAUTICAL RADIONAVIGATION. US93

Non-Gov: AERONAUTICAL RADIONAVIGATION. US93

This band contains the VHF Omnidirectional Range (VOR) and Instrument Landing System (ILS), used worldwide in air navigation. An aircraft receiver computes a bearing to the VOR transmitter from the characteristics of its specially modulated and polarized signal.

Often co-located with VOR but operating in 960–1215 MHz is Distance Measuring Equipment (DME) that provides pilots with distance to the DME station.

As an aid to landing aircraft the ILS incorporates a localizer at 108.1–111.95 MHz, typically 1000 feet beyond the stop end of the runway; a glide slope transmitter at 328.6–335.4 MHz, 1000 feet from the approach end of the runway; and 75 MHz marker beacons that identify positions along the path to the runway.

Aviation users are sensitive to the location, power and interference potential of FM radio stations because of the proximity of the 108–117.975 MHz band to the 88–108 MHz FM broadcast band.

The 112–118 MHz spectrum augments the Global Positioning System (see 1215–1240 MHz) under Special Category I precision approach and landing procedures.

The 108–121.94 MHz spectrum is one of many restricted bands in which the FCC's Part 15 rules permit unlicensed devices to emit only very low level emissions.

## 117.975-121.9375 MHz

Gov: AERONAUTICAL MOBILE (R). 501 591 592 593 US26 US28

Non-Gov: AERONAUTICAL MOBILE (R). 501 591 592 593 US26 US28

FCC: AVIATION (87).

Air traffic controllers use this band to communicate with aircraft, aeronautical utility mobile stations use it for communications among vehicles under tower direction at an airport, and test stations use it to calibrate navigation aids. Federal government aeronautical mobile operations use this and the 225–400 MHz bands.

Industry and government aviation users are pursuing advanced technology in 118–137 MHz, including digital voice and data systems for air traffic services and operational control that will offer considerably more capacity than current systems.

Emergency Position Indicating Radio Beacons (EPIRBs) carried on ships and Emergency Locator Transmitters (ELTs) carried on aircraft transmit distress signals on 121.5 and 243 MHz. Satellites in the international COSPAS/SARSAT search and rescue satellite system monitor these distress signals. The U.S. spacecraft in the system include the GOES and POES weather satellites (see 1670–1690 and 1690–1700 MHz).

The 108–121.94 MHz spectrum is one of many restricted bands in which the FCC's Part 15 rules permit unlicensed devices to emit only very low level emissions.

## 121.9375-123.0875 MHz

Gov: 591 US30 US31 US33 US80 US102 US213

Non-Gov: AERONAUTICAL MOBILE. 591 US30 US31 US33 US80 US102 US213

FCC: AVIATION (87).

This band serves many types of aviation radio, including flight service communication with private and corporate aircraft, weather data transmission, mobile units at airports and flight testing.

Aeronautical advisory stations (unicoms) in this band provide information on local conditions, fuel availability and ground services. Aeronautical search and rescue stations use the frequency 122.9 MHz for training missions.

The 108–121.94 MHz spectrum is one of many restricted bands in which the FCC's Part 15 rules permit unlicensed devices to emit only very low level emissions.

## 123.0875-123.5875 MHz

Gov: AERONAUTICAL MOBILE. 591 593 US32 US33 US112

Non-Gov: AERONAUTICAL MOBILE. 591 593 US32 US33 US112

FCC: AVIATION (87).

Aeronautical flight testing stations use this band. Aeronautical search and rescue stations use 123.1 MHz during emergency missions.

The 123–138 MHz spectrum is one of many restricted bands in which the FCC's Part 15 rules permit unlicensed devices to emit only very low level emissions.

## 123.5875-128.8125 MHz

Gov: AERONAUTICAL MOBILE (R). 591 US26

Non-Gov: AERONAUTICAL MOBILE (R). 591 US26

FCC: AVIATION (87).

This is one of several VHF air traffic control bands.

The 123–138 MHz spectrum is one of many restricted bands in which the FCC's Part 15 rules permit unlicensed devices to emit only very low level emissions.

## 128.8125-132.0125 MHz

Gov: 591

Non-Gov: AERONAUTICAL MOBILE (R). 591

FCC: AVIATION (87).

Aeronautical enroute stations use this band in the Aircraft Communications Addressing and Reporting System

(ACARS). This data network is operated by Aeronautical Radio Inc., a service provider established by the airlines.

ACARS conveys critical information such as the time of aircraft gate departure, liftoff and landing; aircraft position; amount of fuel and weather conditions. It also provides alphanumeric message service for aircrews, controllers and airline offices.

The 123–138 MHz spectrum is one of many restricted bands in which the FCC's Part 15 rules permit unlicensed devices to emit only very low level emissions.

## 132.0125-136 MHz

Gov: AERONAUTICAL MOBILE (R). 591 US26

Non-Gov: AERONAUTICAL MOBILE (R). 591 US26

FCC: AVIATION (87).

This is one of several VHF air traffic control bands. The 123–138 MHz spectrum is one of many restricted bands in which the FCC's Part 15 rules permit unlicensed devices to emit only very low level emissions.

## 136-137 MHz

Gov: 591 US244

Non-Gov: AERONAUTICAL MOBILE (R). 591 US244

FCC: AVIATION (87). SATELLITE COMMUNICATIONS (25).

Air traffic control and domestic and international aviation enroute communications use this band (see also 128.8125–132.0125 MHz).

The 123–138 MHz spectrum is one of many restricted bands in which the FCC's Part 15 rules permit unlicensed devices to emit only very low level emissions.

# 137-137.025 MHz

Gov: SPACE OPERATION (space-to-Earth). METEOROLOGICAL-SATELLITE (space-to-Earth). SPACE RESEARCH (space-to-Earth). MOBILE-SATELLITE (space-to-Earth) 599A 599B US318 US319 US320

Non-Gov: SPACE OPERATION (space-to-Earth). METEOROLOGICAL-SATELLITE (space-to-Earth). SPACE RESEARCH (space-to-Earth). MOBILE-SATELLITE (space-to-Earth) 599A 599B US318 US319 US320

FCC: SATELLITE COMMUNICATIONS (25)

The former 137–138 MHz band has been divided into four bands allocated to the Non-Voice Non-Geostationary (NVNG), low-Earth orbit ("Little LEO") Mobile Satellite Service, which uses small satellites at 650–1300 miles altitude to provide data communication services.

The NVNG service is based on small satellite technology pioneered by the Amateur Radio Service. About 13 million commercial NVNG user terminals are expected to be in use by the year 2000.

The primary NVNG downlink spectrum is 137–137.025, 137.175–137.825 and 400.15–401 MHz. The secondary downlinks are 137.025–137.175 and 137.825–138 MHz (a 387–390 MHz secondary downlink has not yet been domestically implemented).

The primary NVNG uplink spectrum is 148–150.05 and 399.9–400.05 MHz, subject to the closing of navigation satellite use in 149.9–150.05 and 399.9–400.05 MHz.

Primary uplinks in 455–456 and 459–460 MHz and a 312–315 MHz secondary uplink have not yet been domestically implemented. Several of the Little LEO downlink and uplink bands have additional users and sharing criteria that the Little LEOs must accommodate.

The first NVNG player to begin revenue service is Orbital Communications Corp. (Orbcomm), a joint venture of Orbital Sciences Corp. and Teleglobe Canada. Its system will provide two-way data communications to users worldwide through as many as 36 small satellites. Using two commercial satellites, the Orbcomm service began operating in February 1996 with customers in the transportation and environmental monitoring fields.

Manufacturers are introducing small communicators for use with Little LEOs, for applications such as electronic mail, emergency rescue, remote industrial control, cargo and vehicle tracking and recovery. The service will be available through resellers in each participating country.

At this writing the FCC has authorized two other entities to launch NVNG services. Starsys Global Positioning Inc. will serve the industrial messaging and tracking markets, especially transportation and remote data collection.

Volunteers in Technical Assistance (VITA) plans a two-satellite system. VITA is a not-for-profit aid organization supported largely by U.S. foreign aid and research funds. VITA earned the first Pioneer's Preference—an early assurance of license grant—from the FCC.

In awarding the preference, the Commission noted that VITA started experimenting with satellite communications in the early 1980s and applied for an experimental license for a more advanced system in 1988, well before either Orbcomm (1990) or Starsys (1991).

"We believe that VITA's efforts have advanced the authorization of the new service...that will provide reliable, low-cost data communications between ground stations located around the world," the FCC said.

The first VITASAT satellite was destroyed in a failed launch. A replacement will be part of an experimental satellite, FAISAT-2v, to be launched by Final Analysis Communication Services.

Applicants waiting for NVNG licenses include CTA Commercial Systems, E-SAT (a subsidiary of EchoStar, operators of the DISH Network DBS service); GE American Communications (also a majority owner of Starsys); Leo One USA; and Final Analysis.

Most NVNG companies target industrial, utility, transportation and environmental monitoring as key markets, but we expect novel uses to emerge. E-SAT plans to use its satellites to convey billing and interactivity data for TV viewers in the Direct Broadcast Satellite service.

Besides the U.S., Australia, Belgium, France, Germany, Italy and Russia are among the countries developing NVNG systems.

The limited spectrum available for Little LEO satellites has prompted great concern in this industry. Additional NVNG spectrum allocations are under investigation and will be pursued at the 1997 World Radio Conference.

The POES weather satellites output video imagery and telemetry in 137–138 MHz (see 1690–1700 MHz). The band contains other meteorology satellites.

The 123–138 MHz spectrum is one of many restricted bands in which the FCC's Part 15 rules permit unlicensed devices to emit only very low level emissions.

## 137.025-137.175 MHz

Gov: SPACE OPERATION (space-to-Earth). METEOROLOGICAL-SATELLITE (space-to-Earth). SPACE RESEARCH (space-to-Earth). Mobile-Satellite (space-to-Earth). 599A 599B US318 US319 US320

Non-Gov: SPACE OPERATION (space-to-Earth). METEOROLOGICAL-SATELLITE (space-to-Earth). SPACE RESEARCH (space-to-Earth). Mobile-Satellite (space-to-Earth). 599A 599B US318 US319 US320

FCC: SATELLITE COMMUNICATIONS (25)

This is a secondary downlink band in the Non-Voice Non-Geostationary Mobile Satellite Service (see 137–137.025 MHz).

The 123–138 MHz spectrum is one of many restricted bands in which the FCC's Part 15 rules permit unlicensed devices to emit only very low level emissions.

## 137.175-137.825 MHz

Gov: SPACE OPERATION (space-to-Earth). METEOROLOGICAL-SATELLITE (space-to-Earth). SPACE RESEARCH (space-to-Earth). MOBILE-SATELLITE (space-to-Earth) 599A 599B US318 US319 US320

Non-Gov: SPACE OPERATION (space-to-Earth). METEOROLOGICAL-SATELLITE (space-to-Earth). SPACE RESEARCH (space-to-Earth). MOBILE-SATELLITE (space-to-Earth) 599A 599B US318 US319 US320

FCC: SATELLITE COMMUNICATIONS (25)

This is a primary downlink band in the Non-Voice Non-Geostationary Mobile Satellite Service (see 137–137.025 MHz).

The 123–138 MHz spectrum is one of many restricted bands in which the FCC's Part 15 rules permit unlicensed devices to emit only very low level emissions.

## 137.825-138 MHz

Gov: SPACE OPERATION (space-to-Earth). METEOROLOGICAL-SATELLITE (space-to-Earth). SPACE RESEARCH (space-to-Earth). Mobile-Satellite (space-to-Earth) 599A 599B US318 US319 US320

Non-Gov: SPACE OPERATION (space-to-Earth). METEOROLOGICAL-SATELLITE (space-to-Earth). SPACE RESEARCH (space-to-Earth). Mobile-Satellite (space-to-Earth) 599A 599B US318 US319 US320

FCC: SATELLITE COMMUNICATIONS (25)

This is a secondary downlink band in the Non-Voice Non-Geostationary Mobile Satellite Service (Little LEO, see 137–137.025 MHz).

The 123–138 MHz spectrum is one of many restricted bands in which the FCC's Part 15 rules permit unlicensed devices to emit only very low level emissions.

## 138-144 MHz

Gov: FIXED. MOBILE. US10 G30

Non-Gov: US10

This band provides channels for military police, base security, air crash, fire and medical, shipboard radio and other fixed and mobile uses by armed services. The Civil Air Patrol and Coast Guard use it for search and rescue.

This band will be reconfigured to narrowband FM channels, as part of a long-term initiative to recover more communications capacity in federal mobile bands.

## 144-146 MHz

Gov: 510

Non-Gov: AMATEUR. AMATEUR-SATELLITE. 510

FCC: AMATEUR (97).

This band is allocated exclusively worldwide to the Amateur Radio Service.

Hams use this band for experimental, satellite, Morse code and single-sideband communications and reflection of signals off the Moon. Data communications via packet radio is a key use of this band.

Amateur packet radio had its debut in Canada in 1978, when experimenters egged on by forward-thinking regulators transmitted the first amateur packet messages. Inspired by that work, the Tucson Amateur Packet Radio

Corp. in Arizona became a phenomenal success selling kits and licensing designs for packet controllers.

A packet controller formats messages and files into "packets" or short transmissions, encapsulating them with addressing information and sending them via a connected transmitter. A form of time-division multiplexing, packet radio enables many users to efficiently share channels.

Today, amateurs use packet radio networks that automatically convey electronic-mail messages, in this and other bands. Amateur packet radio continues to evolve toward higher speeds, better networks, and systems capable of digital speech and image communication.

## 146-148 MHz

Gov: None

Non-Gov: AMATEUR.

FCC: AMATEUR (97).

Portable and mobile amateur communications in the 146–148 MHz band exploded in the 1970s when American and Japanese manufacturers introduced affordably-priced FM amateur radios. This prompted radio clubs to install repeater stations.

Repeaters provide metropolitan-area mobile radio and telephone service to amateur operators, for little or no outlay of fees other than club dues. Volunteer technical groups coordinate repeater frequencies in Amateur Radio.

## 148-149.9 MHz

Gov: FIXED. MOBILE. MOBILE-SATELLITE (Earth-to-space) 599B US319 US320 US323 US325 608 608A US10 G30

Non-Gov: MOBILE-SATELLITE (Earth-to-space) 599B US319 US320 US323 US325 608 608A US10

FCC: SATELLITE COMMUNICATION (25).

In 1993 the FCC allocated 148–150.05 MHz to the Non-Voice Non-Geostationary Mobile Satellite service (see 137–137.025 MHz). Little LEO licensees must coordinate their systems to minimize interference to existing users.

The 148–149.9 MHz band contains other satellites, government and non-government fixed and mobile uses. POES weather satellites receive commands in this band (see 1690–1700 MHz). This is another band used by the Civil Air Patrol and Coast Guard for search and rescue.

This federal land mobile band is used for many agency missions including law enforcement, transportation, natural resource protection, emergency response, utility operation and for tactical aviation communications.

Scientific data collection such as wildlife telemetry also is conducted in this band. Animals as exotic as gila monsters and sea turtles are outfitted with small transmitters and tracked by researchers. Most satellite-based wildlife telemetry is conducted via the ARGOS system in the 401–402 MHz band.

The main federal land mobile bands are 162.0125–173.2, 173.4–174 and 406.1–420 MHz.

## *149.9-150.05 MHz*

Gov: RADIONAVIGATION-SATELLITE. MOBILE-SATELLITE (Earth-to-space) 599B US319 US322 608B 609A

Non-Gov: RADIONAVIGATION-SATELLITE. MOBILE-SATELLITE (Earth-to-space) 599B US319 US322 608B 609A

The Navy's obsolescent Transit satellites use this band to provide position-location services to ground receivers. Transit and related spacecraft in this spectrum were first launched in 1960 to support submarines. Non-military users include commercial shipping and drilling platforms.

Transit also transmits at 400 MHz. Only one of the two bands is necessary to determine position, but a user gets more accuracy with both. A receiver finds its position by measuring the frequency shift of the satellite signals.

Transit has been superseded by the Global Positioning System (see 1215–1240 MHz). The Navy expects to end Transit service by January 1, 1997. This will clear the way for the Non-Voice Non-Geostationary Mobile Satellite Service (Little LEO) primary VHF uplink spectrum to extend from 148 to 150.05 MHz (see 137–137.025 MHz).

The former Soviet Union has operated Transit-like navigation satellites in this band for decades.

This is one of many restricted bands in which the FCC's Part 15 rules permit unlicensed devices to emit only very low level emissions.

## 150.05-150.8 MHz

Gov: FIXED. MOBILE. US216 G30

Non-Gov: US216

Army, Air Force and Navy systems principally use this band, as does the non-government Emergency Medical Radio Service.

## 150.8-152 MHz

Gov: 613 US216

Non-Gov: LAND MOBILE. 613 US216 NG4 NG51 NG112 NG117 NG124 NG148

FCC: PRIVATE LAND MOBILE (90).

This band contains channels allocated to several Private Mobile Radio Services. Under the FCC's refarming scheme, PMRS channels in 150–174 MHz will transition from the current 15 kHz bandwidth to 6.25 kHz, spaced every 7.5 kHz (see 30.56–32 MHz).

This band is one of several that contain informally standardized "color dot" frequencies used by inexpensive radios. The easy availability of these standardized units contributes to the epidemic of unlicensed operation in the VHF and UHF PMRS bands.

To help combat the problem, the FCC proposed to require an Automatic Transmitter Identification System (ATIS) in mobile and portable radios. The FCC dumped the idea of requiring ATIS in PMRS after trade associations howled.

The Commission decided, however, to require ATIS on satellite uplinks. The need for uplink ATIS became obvious after the unauthorized "Captain Midnight" satellite uplink transmissions in 1986, and the FCC's discovery that erroneous, multiple illuminations of satellites by uplink stations was a common occurrence.

## 152-152.255 MHz

Gov: 613 US216

Non-Gov: LAND MOBILE. 613 US216 NG4 NG51 NG112 NG117 NG124 NG148

FCC: DOMESTIC PUBLIC LAND MOBILE (22).

Public Mobile and public-safety paging operations use this band. It also is used for base stations using the VHF version of Improved Mobile Telephone Service (IMTS) technology that preceded cellular mobile telephone, as well as non-IMTS systems in the Public Mobile Paging and Radiotelephone Service, especially in rural areas. Base stations cover 152.03–152.81 MHz and mobiles 158.49–158.07 MHz (see 454–455 and 459–460 MHz).

Conventional Rural Radiotelephone Stations and Basic Exchange Telephone Radio Systems also use these channels, providing Rural Radiotelephone Service to subscribers not reached by wireline phone service.

The FCC, examining future allocations for Public Safety Radio Services, focused on several bands including 152–152.255, 152.495–152.855 and 157.755–158.115 MHz.

"VHF channels [in those bands] presently allocated for the Public Mobile Service may be lightly used or even unused in some regions," the Commission said. "We believe that this light use of frequencies may result from the discontinuation of service by commercial providers who are intended users of that spectrum." Accordingly, the FCC sought public comment on the feasibility of using those bands for public safety communications.

## 152.255-152.495 MHz

Gov: 613 US216

Non-Gov: LAND MOBILE, 613 US216 NG4 NG51 NG112 NG117 NG124 NG148

FCC: PRIVATE LAND MOBILE (90).

In addition to Business, Forest Products, Taxicab, Special Industrial Radio Service and Public Mobile telephone frequencies, this band contains VHF frequencies for Private Carrier Paging Systems (PCPS).

The FCC established PCPS to provide paging to business users eligible in the Private Mobile Radio Services. Eventually the FCC allowed PCPS operators to provide service to the public on virtually the same unrestricted basis as Public Mobile (common carrier) paging operators. UHF PCPS operations are in the 460 and 929 MHz bands.

PCPS and common carrier paging are considered Commercial Mobile Radio Services (CMRS). The FCC will institute a system of geographic licensing areas to facilitate license auctions in CMRS paging.

The Personal Communications Industry Association predicts that the U.S. paging industry could grow to as many as 56.2 million subscribers by the year 2000.

## 152.495-152.855 MHz

Gov: 613 US216

Non-Gov: LAND MOBILE. 613 US216 NG4 NG51 NG112 NG117 NG124 NG148

FCC: DOMESTIC PUBLIC LAND MOBILE (22).

Public Mobile Paging and Radiotelephone services use this band. It is one of several bands cited by the FCC for possible underutilization (see 152–152.255 MHz).

## 152.855-156.2475 MHz

Gov: 613 US216

Non-Gov: LAND MOBILE. 613 US216 NG4 NG51 NG112 NG117 NG124 NG148

FCC: PRIVATE LAND MOBILE (90). AUXILIARY BROADCASTING (74). MARITIME (80).

A variety of mobile users use this band, including radio and TV Broadcast Auxiliary stations for on-scene reporting, maritime, law enforcement and Private Carrier Paging Systems (see 152.255–152.495 MHz).

## 156.2475-157.0375 MHz

Gov: 613 613A US77 US106 US107 US266

Non-Gov: MARITIME MOBILE. 613 613A US77 US106 US107 US266 NG117

This maritime VHF band contains channels for intership safety, commercial and non-commercial vessel communications, bridge-to-bridge navigation, and state and local control of maritime traffic. These communications increasingly use digital selective calling for semi-automatic call set-up between suitably equipped stations.

In certain port areas the Coast Guard uses maritime channels 11, 12 and 14 for Vessel Traffic Services (VTS). Vessels report their position, navigation and conditions information to the Coast Guard, which tracks the vessels' movements, preventing collisions and groundings. The Coast Guard requires certain large ships, tow and tug boats, dredges and floating platforms to use VTS.

Frequencies allotted for VTS are available for assignment outside of VTS designated areas for other purposes, on a noninterference basis.

Channel 16 (156.8 MHz) is the international distress, safety and calling frequency. Channel 15 (156.75 MHz) is a receive-only channel used to broadcast environmental and safety communications.

The Class C Emergency Position Indicating Radio Beacon (EPIRB), for use by boats in coastal waters, transmits an alert on channel 16 to call attention to a distress situation. It then transmits a longer signal on channel 15 as an aid to direction finding.

The segments 156.52475–156.52525 MHz and 156.7–156.9 MHz are two of the restricted bands in which the FCC's Part 15 rules permit unlicensed devices to emit only very low level emissions.

## 157.0375–157.1875 MHz

Gov: MARITIME MOBILE. 613 US214 US266 G109

Non-Gov: 613 US214 US266

FCC: Private Land Mobile (90).

This is the exclusive federal maritime band. It supports the Coast Guard National Distress and Safety System, which broadcasts safety information on 157.1 MHz.

## 157.1875-157.45 MHz

Gov: 613 US223 US266

Non-Gov: MARITIME MOBILE. 613 US223 US266 NG111

FCC: MARITIME (80).

Ships transmit in this band to maritime public coast stations that provide them with telephone service. The coast stations transmit in 161.8–162.025 MHz.

Industrial/Land Transportation users can use this band away from co-channel public coast stations and waterways.

## 157.45-157.755 MHz

Gov: 613 US266

Non-Gov: LAND MOBILE. 613 US266 NG6 NG28 NG70 NG111 NG112 NG124 NG148

FCC: PRIVATE LAND MOBILE (90).

This band contains Business, Forest Products, Taxicab, Special Industrial Radio Service, Special Emergency Radio Service, Police and Private Carrier Paging frequencies (see 152.255–152.495 MHz).

## 157.755-158.115 MHz

Gov: 613 US266

Non-Gov: LAND MOBILE. 613 US266 NG6 NG28 NG70 NG111 NG112 NG124 NG148

FCC: DOMESTIC PUBLIC LAND MOBILE (22).

This band serves Public Mobile Paging and Radiotelephone Service stations, including Improved Mobile Telephone Service VHF mobile units. It is paired with spectrum for base stations at 152.03–152.81 MHz (see also 454–455 and 459–460 MHz).

These channels also are used by Conventional Rural Radiotelephone Stations and the more modern Basic Exchange Telephone Radio Systems, providing Rural Radiotelephone Service to subscribers not reached by wireline phone service.

This is one of several Public Mobile bands cited by the FCC for possible underutilization (see 152–152.255 MHz).

## 158.115-161.575 MHz

Gov: 613 US266

Non-Gov: LAND MOBILE. (Proposed: MARITIME MOBILE.) 613 NG6 NG28 NG70 NG112 NG124 NG148

FCC: DOMESTIC PUBLIC LAND MOBILE (22).

The FCC defines this as a Public Mobile band, but it serves many different private and public mobile users. These include railroads and power utilities, police radio, highway maintenance, forestry, Broadcast Auxiliary and special frequencies used in oil spill cleanup operations.

It includes mobile transmit frequencies for high-band IMTS telephone systems, and Private Carrier Paging frequencies (see 152.255–152.495 MHz).

The FCC proposed to allocate frequencies in this band for use between public coast stations and ships, and for intership, low power use.

## 161.575-161.625 MHz

Gov: 613 US77

Non-Gov: MARITIME MOBILE. 613 US77 NG6 NG17

FCC: DOMESTIC PUBLIC LAND MOBILE (22). MARITIME (80).

Ships use this band to communicate with coast stations when navigating ports, locks and waterways.

## 161.625-161.775 MHz

Gov: 613

Non-Gov: LAND MOBILE. 613 NG6

FCC: AUXILIARY BROADCASTING (74). DOMESTIC PUBLIC LAND MOBILE (22).

Radio and TV remote pickup stations use this band for on-scene reporting.

## 161.775-162.0125 MHz

Gov: 613 US266

Non-Gov: MARITIME MOBILE. 613 US266 NG6

FCC: DOMESTIC PUBLIC LAND MOBILE (22). MARITIME (80).

Public coast stations use this band to provide telephone service to ships transmitting in 157.2–157.425 MHz.

Industrial/Land Transportation users can operate base stations on frequencies in this band in areas away from co-channel public coast stations and waterways.

## 162.0125-173.2 MHz

Gov: FIXED. MOBILE. 613 US8 US11 US13 US216 US223 US300 US312 G5

Non-Gov: 613 US8 US11 US13 US216 US223 US300 US312

FCC: Auxiliary Broadcasting (74). Private Land Mobile (90).

The main federal VHF mobile bands are 162.0125–173.2 and 173.4–174 MHz. They include some fixed, aeronautical, maritime and non-government uses such as broadcast remote pickup links.

These bands serve many agency missions including federal law enforcement, Presidential protection, building security, transportation, emergency response, utility operation, weather broadcasts and scientific data collection.

These bands have more frequency assignments than any other federal band used for land mobile radio. This band will be converted to narrower channels in order to recover more communications capacity.

The FBI is a principal federal mobile VHF user. In addition to agent-to-agent communications, the FBI—and other agencies including the Drug Enforcement Administration, the Customs Service and the Bureau of Alcohol, Tobacco and Firearms—use this spectrum for audio bugging devices, tracking beacons, body transmitters and control of information collection equipment.

Some of these uses were impacted by a 1989 FCC and NTIA decision to allocate a key FBI channel, 173.075 MHz, to the Stolen Vehicle Recovery Service (FCC Rule 90.19(f)(5)(7)). The sole user of this channel is Lo-Jack Corp. which sells tracking beacons to consumers.

Lo-Jack claims impressive success in vehicle recovery. Installers hide the beacons inside vehicles. When a car is reported stolen, computers cause the Lo-Jack network to activate the beacon. Direction-finder patrol cars can then locate the vehicle. Lo-Jack equipped police cars are easy to identify by their four vertical antennas.

The U.S. Attorney General and the FBI Director of the fought reallocation of this channel. The matter even surfaced in Congressional testimony. A FBI witness told a 1990 House Telecommunications Subcommittee panel: "The value of one of our few nationwide channels, used to service body transmitters and tracking beacons, was seriously and adversely diminished, and our operational capabilities adversely affected when a private sector firm was granted nationwide use of this channel."

Lo-Jack and VHF vehicle location should not be confused with the Location and Monitoring Service at 902–

928 MHz or with covert tracking devices used by police in 30–50 MHz. The Location and Monitoring Service, though an older service historically, now uses modern spread spectrum technology and newer methods of pinpointing vehicles and other beacon-equipped objects.

Spectrum efficiency improvement plans of the National Telecommunications and Information Administration require that all new equipment in 162–174 MHz must be capable of operating within a 12.5 kHz–wide channel. By January 1, 2005, all equipment must do so. The 406.1–420 and 138–150.8 MHz federal bands also will be rechanneled to narrower channel spacings. All equipment operating within these bands will be converted by 2008.

The National Oceanic and Atmospheric Administration's National Weather Service operates NOAA Weather Radio on 162.400, 162.425, 162.450, 162.475, 162.500, 162.525 and 162.550 MHz. The signal may be received on scanners and on special radios sold by major electronics dealers. Broadcasting from hundreds of transmitters across the country, NOAA Weather Radio provides forecasts, warnings and related information to the public.

The digital version of NOAA Weather Radio is EMWIN, the Emergency Managers Weather Information Network, broadcast experimentally in this band. EMWIN transmissions include graphics for display on computers using inexpensive hardware and software. The GOES 8 satellite also broadcasts EMWIN (see 1690–1700 MHz).

A new government initiative aims to turn NOAA Weather Radio into a so-called "all hazards" broadcaster, to expand its reach to most of the nation, and to make its receivers "as common as smoke detectors in homes," according to Vice President Al Gore.

This program to make emergency radios ubiquitous is not the first. When NOAA Weather Radio was established in the 1970s, it was proclaimed to be the successor to the ill-fated Public Emergency Radio, internally called the Decision Information Distribution System (DIDS).

Though not in SpectrumGuide's 30 MHz–300 GHz range, DIDS warrants examination as a historical oddity. Operating in 160–190 kHz, DIDS was an ambitious scheme to use longwave to warn the public of attack. Like Weather Radio, it was to use special receivers in every home.

The $2 million flagship DIDS station, WGU–20 in Chase, Md., operated from 1972 to 1984, broadcasting mysterious announcements. It had a 700-foot antenna and was the first all solid-state 50 kW AM station.

DIDS was declared the successor to CONELRAD (CONtrol of ELectromagnetic RADiation), a 1960s defense program that required AM stations to turn on and off and vary their schedules.

They were supposed to warn the public while confusing missiles that might home in on broadcast signals. Triangular CONELRAD logos were printed on the dials of radio receivers of this period. CONELRAD was discontinued when missile technology improved.

DIDS' creators regarded existing warning technology as prone to human error and delay. DIDS was to be a spoof-proof supersystem that could switch on sirens and convey news to officials by radioteletype. DIDS' unique selling proposition was that it would speak to citizens in their cars and homes, even waking them from sleep.

Most Americans do not own longwave receivers, however. "The acquisition of home warning receivers would be a voluntary decision on the part of the individual citizen," said one DIDS manual. Thus, the government had to persuade manufacturers to market and the public to buy radios whose only use would be in national emergencies.

Westinghouse developed prototype receivers. One of them could attach to, or be installed inside, a TV set. If the

TV was turned off when DIDS was activated, the receiver would deliver the message at high volume through the TV speaker to wake sleepers.

To create an approachable image for consumers, DIDS was renamed Public Emergency Radio (PER) and given a cartoon puppy mascot named PERKI.

The U.S. antiballistic missile system was designed to connect directly to PER. Radar could track the flight paths of incoming enemy missiles. While the radar network dispatched ABMs to destroy the missiles at high altitudes, it would instantly furnish a prediction to PER listeners as to where an enemy warhead would land and explode if it was not successfully intercepted by an ABM. The PER broadcast would then tell commuters which areas to avoid.

Federal Emergency Management Agency officials told us that they discontinued PER because of the high cost to build and operate enough stations to blanket the U.S. It also would have been a one-way system, without a means to obtain status reports and damage assessments. WGU-20 was eventually dismantled.

The principal electronic connection between federal, state and local emergency officials is the National Warning System, a landline telephone conferencing network admittedly subject to disruption by disasters.

If not by NOAA Weather Radio, most Americans would probably receive military emergency warnings over the Emergency Alert System (EAS, see 88–108 MHz).

The segments 162.0125–167.17 and 167.72–173.2 MHz are restricted bands in which the FCC's Part 15 rules permit unlicensed devices to emit only very low level emissions.

## 173.2-173.4 MHz

Gov: None

Non-Gov: FIXED. Land Mobile. NG124

FCC: Private Land Mobile (90).

Power, Forest Products, Special Industrial, Petroleum, Manufacturers, Business and Local Government Radio Services use this band for telemetry and remote control.

Except for frequencies near the edges, frequencies in this band are spaced 12.5 kHz apart. In designing its refarming scheme to obtain more capacity in the Private Mobile Radio Services (see 30.56–32 MHz), the FCC decided not to change the channel bandwidth arrangements in this band due to the heavy use and unique channel spacing.

## 173.4-174 MHz

Gov: FIXED. MOBILE. G5

Non-Gov: None

This is part of the federal VHF spectrum mostly used for land mobile communications, with some fixed, aeronautical and maritime mobile uses (see 162.0125–173.2 MHz).

## 174-216 MHz

Gov: None

Non-Gov: BROADCASTING. NG115 NG128 NG149

FCC: RADIO BROADCAST (TV) (73). Auxiliary Broadcasting (74).

This band contains TV channels 7 through 13.

The FCC proposed to relocate all U.S. TV broadcasting to channels 7–51 in the transition to Digital Television (see 470–512 MHz).

Wireless microphones and telemetry devices such as electrocardiogram monitors are permitted in this band. The FCC restricts power levels of these devices to protect TV reception. Telemetry manufacturers argued that this band is overcrowded and affected by a high level of ambient electrical noise in healthcare facilities.

In response, the FCC proposed to increase the power levels permitted to biomedical telemetry and to require that the devices be separated from TV stations by certain distances. The FCC also proposed to give these devices access to UHF TV channels (470–608 and 614–806 MHz).

## 216-220 MHz

Gov: MARITIME MOBILE. Aeronautical Mobile. Fixed. Land Mobile. Radiolocation. 627 US210 US229 US274 US317 G2

Non-Gov: MARITIME MOBILE. Aeronautical-Mobile. Fixed. Land Mobile. 627 US210 US229 US274 US317 NG152

FCC: MARITIME (80). Private Land Mobile (90). Personal Radio Service (95). Amateur (97).

This band is home to diverse transportation, military and personal radio services.

Automated Maritime Telecommunication Systems (AMTS) use this band for ship telephones. AMTS users can dial the number of the desired vessel or landline phone. The system connects the call and records billing information.

Most AMTS stations are along or near the Mississippi River, Ohio River and the Gulf Coast, but the FCC permits them anywhere in the U.S. AMTS network control is accomplished in the Low Power Radio Service (see below).

This band is used in geophysical exploration. Seismic telemetry uses a sound source such as a dynamite charge that introduces seismic energy into the Earth's surface. Geophones pick up the reflected energy, and convey the resulting data by radio to receivers and recorders.

Among several telescope and radar systems, the Naval Space Command (NAVSPACECOM) operates a colossal satellite and missile-sensing radar on 216.98 MHz. Formerly known as the "NAVSPASUR (Naval Space Surveillance) Fence"—and now simply the "Fence" or the "Sensor"—it uses a main transmitter at Lake Kickapoo/Archer City, Texas and subsidiary transmitters at Jordan Lake, Alabama and Gila River, Arizona.

Energy from this multibillion-watt system forms an electromagnetic fence extending 3000 miles from California to Georgia, 1000 miles off each coast and thousands of miles into space.

Aircraft and spacecraft passing through the beam reflect the energy back to Earth, where it is picked up by receivers in San Diego, Calif.; Elephant Butte, N.Mex.; Red River, Ark.; Silver Lake, Miss.; and Tatnall and Hawkinsville, Ga.

The receivers measure the reflected signals—more than one million per month—and convey the data to NAVSPACECOM headquarters in Dahlgren, Va. This facility updates a database of spacecraft orbital elements and informs the U.S. Space Command Space Surveillance Center (SSC) in Cheyenne Mountain, Colorado.

NAVSPACECOM is the alternate SSC. It performs computer analysis to predict collisions of spacecraft with each other and with spaceborne debris. The Fence can track space objects at least 30 cm. (approx. 1') in size.

It also detects anomalous, uncorrelated targets, as well as a category reportedly designated "Alternative Space Craft" (ASC). UFO investigators have tried unsuccessfully to gain access to the ASC database.

The 218–219 MHz segment is allocated to the Interactive Video and Data Service (IVDS). Although the FCC classifies IVDS as a Personal Radio Service (like Citizens Band and the General Mobile Radio Service), it actually is a commercial point-to-multipoint and multipoint-to-point fixed service. The FCC also permits mobile IVDS.

EON Corp. (formerly TV Answer) conceived IVDS as an interactive service that would permit home shopping. EON invested hundreds of millions of dollars in IVDS, later turning to wireless E-mail and vertical data markets when the interactive TV market failed to emerge.

"Depending on what the licensee chooses to offer, IVDS could combine the functions of computers, TV sets and compact-disk players," according to the FCC. "Subscribers will be able to choose the camera angle during a sporting event, pay bills, shop-until-they-drop at malls, choose endings to TV shows, check college catalogs, play video games, choose movies on demand, or order a pizza with or without toppings."

A user device, the Response Transmitter Unit (RTU) forwards the user's identity and menu selections to the nearest IVDS licensee Cell Transmitter Station (CTS) and network center, which forwards the data via satellite to EON's service bureau.

The FCC grants two IVDS licenses, 500 kHz each, for each of 734 service areas. The FCC granted licenses in subsequent markets by auction, but a lottery awarded the first IVDS licenses in New York City, Los Angeles, Chicago, Philadelphia, Boston, San Francisco, Washington D.C., Dallas and Houston.

The FCC IVDS technical rules do not actually require that the IVDS licensee connect to or utilize television.

At EON's request, the FCC decided to allow mobile RTUs in the IVDS. "This new [mobile] technology will allow IVDS licensees to leverage existing platforms such as the new generation of portable game players, portable televisions and personal organizers in order to provide customers and program providers greater options to customize and better organize the data that they transmit over the IVDS system," EON said.

"At price points similar to basic local paging services, parents will be able to confirm their child's safety or send an electronic 'come home now,'" EON added. "Broadcasters will be able to provide sports updates to people on the road and let them know to tune in at 7 PM when the evening game begins."

EON's patent (5,388,101) states that its RTUs may be "put into a car or van for movement within or across cell boundaries with good digital synchronous communication contact within the nationwide network of cells." This broad claim appears to describe any digital cellular mobile communication system, but EON maintains that others may use technology that complies with FCC IVDS rules without infringing the patent.

Initially rejecting the idea of mobile IVDS use, the FCC said that "The primary use of the IVDS system must be to provide subscribers at fixed locations with the capability to interact with video, data or other service providers. The offering to subscribers of mobile service only, such as paging or dispatch services, would not be permitted."

The FCC later concluded differently. "Commenters favoring mobility have stated that IVDS spectrum will be useful for TV-based consumer messaging, as well as for messaging in the business and commercial data context, with applications for meter reading, stock transactions or quotations on demand, and automatic teller machine and credit card verifications. ...

"To the extent that there is public demand for mobile IVDS, our action here [allowing mobile IVDS] will permit licensees to respond to this demand. This flexibility should also encourage a broader range of service offerings."

Mobile RTUs are limited to 100 milliwatts effective radiated power. Fixed RTUs are permitted up to 20 watts.

Fixed RTUs must automatically select the minimum power necessary if they use more than 100 milliwatts.

To protect against interference to TV channel 13, the FCC strictly limits the duration of transmission from each RTU in areas served by TV stations on that channel.

IVDS licenses "are presently unusable and have no clearly identifiable use on the near or distant horizon," according to Total Interaction Inc., holder of several major market IVDS licenses. At this writing, no IVDS systems are known to be in revenue operation.

The Amateur Radio Service was allocated 219–220 MHz on a secondary basis for digital network links. The American Radio Relay League (ARRL) requested the allocation as compensation for the loss of amateur access to 220–222 MHz.

Amateur use in 219–220 MHz is limited to directional fixed stations so as not to interfere with AMTS. Operations must be coordinated with AMTS licensees.

High data rate networks in this band allow amateurs to send audio and video data as well as the text messages and files common today in the amateur packet system. Mobile uses and point-to-multipoint operations, typical of packet radio base stations, are not allowed in this band. The FCC currently is considering whether to permit spread spectrum amateur transmissions in this band as well.

This band also is home to another fascinating application: tracking of stolen currency and merchandise. The current sole provider of this service is ProNet, a paging carrier. Its electronic tracking subsidiary ETS operated for more than 20 years under FCC experimental authority, apparently one of the longest running experiments the agency has licensed.

Banks and retailers hide the ETS "TracPac" transmitter in paper money or other valuables. "Once the TracPacs are activated by a crime, highly sensitive directional receivers, which are located in law enforcement vehicles, helicopters, or at a few fixed sites, track and locate the crime while it is still in progress," the company said. "Law enforcement directly monitors for those signals in its own dispatch center, and when triggered, previously deployed trackers scramble to locate the unlawful incident, protect the public, and apprehend the felon. On average, all this occurs in less than 15 minutes."

ETS has an impressive record of arrests and convictions in armed robberies, theft of automatic teller machines, and other serious crimes. The company is expanding into panic alarm transmitters for use on college campuses.

The FCC eventually granted this system a spectrum home in the Low Power Radio Service (LPRS) in 216–217 MHz.

Other LPRS channels are devoted to auditory assistance devices for the hearing-impaired and a broad category of "health care assistance devices".

Auditory assistance devices consist of a short range transmitter and headset receivers for use in classrooms, churches or theaters by persons with hearing disabilities (see 72–73 MHz). Interference from other uses of the 72 MHz spectrum prompted manufacturers to request spectrum in the 216–217 MHz band for these products.

"Unfortunately, there is no way to predict what explosions of technology may occur and lead to unexpected interference sources in this band, in a manner that is similar to what occurred in the 72–76 MHz band with the advent of hundreds of paging companies," according to Karen Strauss of the National Association of the Deaf.

The FCC also allows Automated Maritime Telecommunications System coast stations to use the LPRS for network control. All LPRS operations are considered Citizens Band Radio Services in Part 95 of the FCC rules. They are not unlicensed devices, but they are "authorized by rule" (no license documents are issued).

# 220-222 MHz

Gov: LAND MOBILE. Radiolocation. 627 G2

Non-Gov: LAND MOBILE. 627

FCC: PRIVATE LAND MOBILE (90)

In 1988 the FCC reallocated 220–222 MHz to the Private Mobile Radio Services from the 220–225 MHz band used by Amateur Radio. The principal requester was United Parcel Service which needed radios for its vehicles.

The FCC received about 60,000 220 MHz license applications in 1991. After Congressional hearings into and years of litigation over the propriety of the FCC's 220 MHz allocation and licensing process, the Commission awarded licenses by lottery in 1993.

The 220–222 MHz band gradually began to develop new narrowband voice and data services. Narrowband channels occupy less bandwidth than conventional channels. These services basically provide dispatch of vehicle fleets and business and maintenance communications.

Through management agreements with license holders, several firms have built 220–222 MHz regional and nationwide networks to serve those markets.

In what the FCC called Phase I, it divided the band into 400 5 kHz–wide channels to create 200 channel pairs. Base stations operate in 220–221 and mobile units in 221–222 MHz. The channel pairs were allotted to several categories of licensees, including commercial (for-profit communications service to customers) and non-commercial (communications for the licensee's internal use).

Some licenses were devoted to local use and others to nationwide coverage; several channels were made available to public safety and emergency medical users.

Phase II of the 220–222 MHz band will have a different technical and regulatory scheme. The FCC believes that this band will mostly be used to provide Commercial Mobile Radio Service (CMRS)—that is, telephone-interconnected mobile service to the public—and thus will essentially be similar to other forms of CMRS such as cellular and Personal Communications Services (PCS) and Specialized Mobile Radio (SMR). Mutually-exclusive applications for licenses in this band are subject to auction.

The FCC plans three types of Phase II 220 MHz licenses: nationwide, regional, and local "EA" (Economic Areas) as defined by the Commerce Department, with several licenses available in each category. Licensees could aggregate channels to take advantage of non-narrowband technologies such as Time-Division Multiple Access. Licensees also could offer fixed and paging services.

The federal government held co-primary allocations on 125 non-nationwide channels in 220–222 MHz, but will relinquish its co-primary status when the FCC adopts its final regulatory scheme for the band. Those channels will become available to non-federal users on a primary basis.

But for an acrimonious defeat, the Personal Emergency Locator Transmitter Service (PELTS), proposed by a cellular engineer, would have operated in this 220 MHz band. PELTS was a portable rescue radio system for use in remote areas. The FCC saw PELTS as an alternative to the illegal operation on land of emergency radio beacons intended for ships and aircraft (see 406–406.1 MHz).

PELTS would have had channels for user contact with watch centers and for one-way broadcasts, alerting and homing, and short-distance voice.

As is often the case with novel spectrum proposals, worried stakeholders attacked PELTS. Emergency rescue groups argued that PELTS would expose them to lawsuits in responding, or failing to respond, to emergency calls.

The FCC's PELTS licensing scheme required consumers to rent the radios from search and rescue teams, and prevented consumers from owning the radios themselves.

Radio amateurs warned that PELTS would become another Citizens Band service, filled with chaotic traffic and false alarms. Instead of a life-saving technology, most commenters painted PELTS as a government gewgaw dreamed up by bureaucrats.

"Based on the comments it does not appear that a watch and response system so critical to the success of PELTS will materialize," the FCC conceded. "Further, this lack of support to promote and develop the technology for PELTS would result in uneven usage and a lack of uniformity of emergency communications service. ...

"The unsettled nature of the liability question and the legal trend of pursuing large damage suits would likely preclude knowledgeable search and rescue individuals and organizations from participating in PELTS."

The FCC's rejection of PELTS appeared to leave Mt. Hood, Oregon's Mountain Signal system in limbo. Its users rent small transmitters from outdoor outfitters. In case of emergency, they can use the transmitters to signal to receivers around the area.

The transmitters are manufactured for, monitored by and licensed experimentally to a not-for-profit corporation. Medical, legal and wilderness experts established the corporation after mountaineering accidents occurred. State legislation gave the corporation legal immunity.

The FCC never issued Mountain Signal permanent rules or regular license and it is not legal to use anywhere else.

The best emergency outdoor communications system will probably be Little LEO and Big LEO satellites (see 137–137.025 and 1610–1610.6 MHz), particularly in areas without PCS or cellular coverage.

## 222-225 MHz

Gov: Radiolocation 627 G2

Non-Gov: AMATEUR. 627

FCC: AMATEUR (97)

The Amateur Radio 1.25 meter band occupies 222–225 MHz. It originally extended from 220 to 225 MHz, but the FCC reallocated the lower 2 MHz to create the commercial narrowband mobile allocation in 220–222 MHz.

## 225-235 MHz

Gov: FIXED. MOBILE. 501 592 642 644 G27 G100

Non-Gov: 501 592 642 644

The 225–400 MHz spectrum is devoted to military aircraft, tactical and training communications, satellite links for ground, air, surface and subsurface users, rocket test and telemetry and Presidential communications.

The FCC called on the government to offer some of this spectrum to the private sector. Other nations use portions of it for non-military purposes (see 335.4–399.9 MHz).

"Spectrum reallocated in the 225–400 MHz band could greatly facilitate our efforts by providing 'green space' in which to begin implementing spectrum-efficient systems," the FCC said. "This spectrum could also provide greater flexibility for accommodating applications that require wider bandwidth, such as transmission of data or images. ...

"Spectrum in this band could also be very valuable in enhancing public safety communications and could possibly provide spectrum for use by biomedical telemetry devices, which manufacturers have indicated require spectrum below 1 GHz."

## 235-273 MHz

Gov: FIXED. MOBILE. 501 592 642 644 G27 G100

Non-Gov: 501 592 642 644

This is part of the 225–400 MHz military spectrum (see 225–235 MHz).

NASA uses frequencies in this part of the spectrum for Space Shuttle spacesuit-orbiter, orbiter-ground and emergency crew return communications.

Also operating in 244–270 MHz is FLTSATCOM, the Navy's Fleet Satellite Communications System connecting aircraft, ships, submarines and command elements. The Air Force AFSATCOM system and the Presidential Command Network have channels on these satellites.

The Navy is replacing its FLTSATCOM, Gapfiller and Leasat satellites with the $1.8 billion UHF Follow-On (UFO) series of satellites. These nuclear-hardened satellites will use the same spectrum but provide more channels.

Some of the UFOs contain packages for the Global Broadcast System (GBS), the Pentagon's version of Direct Broadcast Satellite (DBS) television for military users.

Each UFO GBS package includes transponders for spot (localized) and broad-area coverage beams, serving small 22-inch receiving dishes. Field users will be able to request video, still images or other data and receive it via GBS in real time. A typical use of GBS is to distribute imagery received by unmanned aerial vehicles.

The newer UFO satellites include Extremely High Frequency (EHF) communications packages compatible with Milstar ground terminals.

Milstar is the Pentagon's enormous UHF/SHF/EHF nuclear-proof satellite project. It has UHF uplinks and downlinks in the 225 MHz spectrum for compatibility with older systems (see 20.2–21.2 GHz for Milstar information).

Also in this band are Emergency Position Indicating Radio Beacons (EPIRBs) carried on ships, and Emergency Locator Transmitters (ELTs) carried on aircraft. There are several types of EPIRBs with different characteristics such as manual activation, automatic activation, survival craft use and coastal use.

These devices transmit distress signals on 121.5 and 243 MHz and the more modern 406.025 MHz in the international COSPAS/SARSAT search and rescue satellite system. U.S. satellites in this system include the GOES and POES weather satellites (see 1670–1690 and 1690–1700 MHz). The coastal Class C EPIRB transmits in the maritime VHF band (see 156.2475–157.0375 MHz).

The spectrum 240–285 MHz is one of many restricted bands in which the FCC's Part 15 rules permit unlicensed devices to emit only very low level emissions.

## 273-312 MHz

Gov: FIXED. MOBILE. 501 592 642 644 G27 G100

Non-Gov: 501 592 642 644

This is part of the 225–400 MHz military spectrum (see 225–235 MHz).

FCC Rules do not prescribe this specific band for unlicensed Part 15 alarms and controls, but the rules do allow them in 40.66–40.70 MHz and 70 MHz–38.6 GHz except for restricted bands limited to safety and critical government operations.

Car alarm and remote control manufacturers often choose this band because this spectrum otherwise contains only federal users. Commonly-used frequencies for these devices are 303.8, 315 and 318 MHz.

That approach is not foolproof. High-power transmissions in this spectrum from Presidential and military aircraft are thought responsible for interference that causes openings of home garage doors in the flight path.

The spectrum 240–285 MHz is one of many restricted bands in which the FCC's Part 15 rules permit unlicensed devices to emit only very low level emissions.

## 312-315 MHz

Gov: FIXED. MOBILE. 501 592 642 644 G27 G100

Non-Gov: 501 592 642 644

This is part of the 225–400 MHz military spectrum (see 225–235 MHz).

The band is supposed to be available to uplinks in the Non-Voice Non-Geostationary Mobile Satellite Service on a secondary basis, but the allocation has not been made at this writing. The companion downlink is 387–390 MHz.

The federal government is extremely sensitive about sharing with non-governmental uses in 225–400 MHz.

## 315-322 MHz

Gov: FIXED. MOBILE. 501 592 642 644 G27 G100

Non-Gov: 501 592 642 644

This is part of the 225–400 MHz military spectrum (see 225–235 MHz).

## 322-328.6 MHz

Gov: FIXED. MOBILE. 501 592 642 644 G27 G100

Non-Gov: 501 592 642 644

This is part of the 225–400 MHz military spectrum (see 225–235 MHz).

The 322–328.6 MHz band, although not allocated to passive (receive-only) services in the U.S., is widely used by radio astronomers for pulsar observations. The spinning pulsars remain after stars have expended all of their fuel and collapsed into dense spheres a few miles in diameter.

Pulsars emit rotating beams of radio energy. The extreme regularity of pulsar emissions appeared so unusual when discovered in 1967 that scientists suspected that they were transmissions from intelligent beings.

The 322–335.4 MHz spectrum is one of many restricted bands in which the FCC's Part 15 rules permit unlicensed devices to emit only very low level emissions.

## 328.6-335.4 MHz

Gov: AERONAUTICAL RADIONAVIGATION. 645

Non-Gov: AERONAUTICAL RADIONAVIGATION. 645

This band is used worldwide for the aircraft Instrument Landing System (see 108–117.975 MHz).

The 322–335.4 MHz spectrum is one of many restricted bands in which the FCC's Part 15 rules permit unlicensed devices to emit only very low level emissions.

## 335.4-399.9 MHz

Gov: FIXED. MOBILE. G27 G100

Non-Gov: None

This is part of the 225–400 MHz military spectrum (see 225–235 MHz).

The FCC has identified this band as a possible home for future Public Safety Radio Services. An agreement between the North American Treaty Organization (NATO) and European government bodies provides for European use of this band for public safety services.

"This low UHF segment, therefore, appears to be a good candidate for reallocation for public safety uses," the FCC has observed. "Its proximity to the 74 channels in the 450–470 MHz band that are currently available for public safety communications may make this segment ideal for public safety voice, data and facsimile channels. We note, however, that these frequencies are regulated by NTIA." The Commission sought public comment on whether and how this band could be reallocated.

The 387–390 MHz segment (for downlinks) is a possible companion to the 312–315 MHz uplink band under consideration for expansion of the Non-Voice Non-Geostationary Mobile Satellite Service (Little LEO).

## *399.9-400.05 MHz*

Gov: RADIONAVIGATION-SATELLITE. MOBILE-SATELLITE (Earth-to-space). 645B US319 US326

Non-Gov: RADIONAVIGATION-SATELLITE. MOBILE-SATELLITE (Earth-to-space). 645B US319 US326

The Non-Voice Non-Geostationary Mobile Satellite Service primary UHF uplink is now allocated this band worldwide (see 137–137.025 MHz).

The Transit navigation satellite system transmits in this band, as does a similar Russian system (see 149.9–150.05 MHz). The Navy is expected to discontinue Transit service by January 1, 1997.

The 399.9–410 MHz spectrum is one of many restricted bands in which the FCC's Part 15 rules permit unlicensed devices to emit only very low level emissions.

## 400.05-400.15 MHz

Gov: STANDARD FREQUENCY AND TIME SIGNAL-SATELLITE. 646

Non-Gov: STANDARD FREQUENCY AND TIME SIGNAL-SATELLITE. 646

The frequency 400.1 MHz is allocated for frequency and time broadcasts from satellites. The allocation was intended to replace some HF time and frequency services. The U.S. does not operate satellites in this band.

The 399.9–410 MHz spectrum is one of many restricted bands in which the FCC's Part 15 rules permit unlicensed devices to emit only very low level emissions.

## 400.15-401 MHz

Gov: METEOROLOGICAL AIDS (radiosonde). METEOROLOGICAL-SATELLITE (space-to-Earth). SPACE RESEARCH (space-to-Earth). MOBILE-SATELLITE (space-to-Earth). Space Operation (space-to-Earth). 599B 647 647A 647B US70 US319 US320 US324

Non-Gov: METEOROLOGICAL AIDS (radiosonde). SPACE RESEARCH (space-to-Earth). MOBILE-SATELLITE (space-to-Earth). Space Operation (space-to-Earth) 599B 647 647A 647B US70 US319 US320 US324

In 1993 the FCC allocated this band to Non-Voice Non-Geostationary mobile satellites (see 137–137.025 MHz).

The 400.15–406 MHz meteorological aids spectrum is used for weather and water data collection by governments and researchers. The Defense Meteorological Satellite Program also uses this band (see 1690–1700 MHz).

This is one of many restricted bands in which the FCC's Part 15 rules permit unlicensed devices to emit only very low level emissions.

# 401-402 MHz

Gov: METEOROLOGICAL AIDS (radiosonde). SPACE OPERATION (space-to-Earth). Earth Exploration-Satellite (Earth-to-space). Meteorological-Satellite (Earth-to-space). US70

Non-Gov: METEOROLOGICAL AIDS (radiosonde). SPACE OPERATION (space-to-Earth). Earth Exploration-Satellite (Earth-to-space). Meteorological-Satellite (Earth-to-space). US70

FCC: SATELLITE COMMUNICATIONS (25).

Radiosonde balloons and other so-called "data platforms" transmit in 401–406 MHz to the NOAA GOES and POES satellites (see 1670–1690 and 1690–1700 MHz).

As coordinated by the World Meteorological Organization, radiosondes worldwide are launched at the same time to provide a global snapshot of air movement, pressure and temperature. This information is fed into computer models to produce weather forecasts.

Akin to radiosondes are dropsondes, instrument packages with GPS or other tracking receivers that are parachuted from aircraft. Dropsondes measure temperature and humidity, and transmit the data to the aircraft.

Data platforms collect meteorological data from sensors in remote locations. They measure stream height, snow depth and temperatures in mountain areas. This data is used to predict droughts and floods. Other sensors on ships and buoys measure ocean temperature and ice floe velocity and direction, and upper atmosphere temperature from aircraft.

Included among these scientific uses are wildlife telemetry and emergency communications for expeditions. These science platform applications use ARGOS, a French/U.S. service at 401.65 MHz.

ARGOS enables location of a transmitter anywhere in the world to within 150–1000 meters. A celebrated ARGOS "customer" is Ning, a Malaysian elephant equipped with a radio necklace. The Smithsonian Institution and the Malaysian National Parks Department track the movement of elephants in studies of their feeding and environment.

The ARGOS data collection system is carried onboard the NOAA POES satellites. The system receives the incoming signal, measures its frequency and time characteristics, and retransmits this data over the POES VHF beacons to collection networks accessible to subscribing scientists.

The Japanese and European space agencies plan a series of spacecraft to join the POES satellites in orbit. They will provide ARGOS with the ability to transmit messages from the satellite to ground units.

Experimental satellite operations focus on 401–404 MHz as a candidate for expansion of the Non-Voice Non-Geostationary Mobile Satellite Service (Little LEO, see 137–137.025 MHz).

This is one of many restricted bands in which the FCC's Part 15 rules permit unlicensed devices to emit only very low level emissions.

## *402-403 MHz*

Gov: METEOROLOGICAL AIDS (radiosonde). Earth Exploration-Satellite (Earth-to-space). Meteorological-Satellite (Earth-to-space). US70

Non-Gov: METEOROLOGICAL AIDS (radiosonde). Earth Exploration-Satellite (Earth-to-space). Meteorological-Satellite (Earth-to-space). US70

See 400.15–401 and 401–402 MHz for uses of this band.

It is one of many restricted bands in which the FCC's Part 15 rules permit unlicensed devices to emit only very low level emissions.

## 403–406 MHz

Gov: METEOROLOGICAL AIDS (radiosonde). US70 G6

Non-Gov: METEOROLOGICAL AIDS (radiosonde). US70

See 400.15–401 and 401–402 MHz for uses of this band.

Wind profilers—weather radars used in aviation and pollution studies—used to operate in this band but were removed because they interfered with COSPAS/SARSAT search and rescue satellites in 406–406.1 MHz.

The FCC proposed to allocate 448–450 MHz to profilers, subject to coordination with military users and notification to repeater station operators in the Amateur Radio Service.

This is one of many restricted bands in which the FCC's Part 15 rules permit unlicensed devices to emit only very low level emissions.

## 406–406.1 MHz

Gov: MOBILE-SATELLITE (Earth-to-space). 649 649A

Non-Gov: MOBILE-SATELLITE (Earth-to-space). 649 649A

Emergency Position Indicating Radio Beacons (EPIRBs) on ships and Emergency Locator Transmitters (ELTs) on aircraft transmit signals in this band that are used to locate and identify craft and persons in distress. This band is protected for this purpose throughout the world.

Older EPIRBs and ELTs operate on 121.5 and 243 MHz. Newer EPIRBs use 406.025 MHz. Transmissions on the older frequencies provide distress alerting and homing assistance but not identification.

The 406.025 MHz units transmit identification data that must be registered with the National Oceanic and Atmospheric Administration. NOAA maintains a database of identifications and provides them to rescue authorities.

COSPAS/SARSAT-equipped satellites monitor EPIRB and ELT signals and alert control centers to start search and rescue efforts. COSPAS is an acronym for a Russian phrase that means “space system for search and distress vessels.” SARSAT stands for Search and Rescue Satellite-Aided Tracking. Twenty-seven nations fund the London-based COSPAS/SARSAT program.

The 121.5 MHz signal can only be detected and relayed when a satellite is in range of the EPIRB or ELT and a ground station called a Local User Terminal (LUT). There are areas where such signals cannot be relayed to a LUT.

In contrast, the 406.025 MHz signal is immediately processed on board the satellite, stored, and transmitted to the ground on 1544.5 MHz from a continuous memory dump. Ground stations acquire the data and beacon identification, and forward the alerts to control stations and rescue centers.

U.S. satellites in the system include the GOES and POES weather satellites (1690–1700 and 1700–1710 MHz). Russia also operates satellites with COSPAS/SARSAT payloads, and several additional nations will launch satellites with this capability. Many rescues are attributed to this system. The market for EPIRB and ELT equipment is brisk.

Misuse of EPIRBs and ELTs on land prompted the FCC to propose a Personal Emergency Locator Transmitter Service (PELTS). Rescue groups denounced PELTS and refused to participate (see 220–222 MHz).

Land-based personal emergency communications will probably best be served by Big LEO and Little LEO satellites (see 137–137.025 and 1610–1610.6 MHz).

This is one of many restricted bands in which the FCC's Part 15 rules permit unlicensed devices to emit only very low level emissions.

# 406.1–410 MHz

Gov: FIXED. MOBILE. RADIO ASTRONOMY. US13 US74 US117 G5 G6

Non-Gov: RADIO ASTRONOMY. US13 US74 US117

The 406.1–410 and 410–420 MHz bands are changing from mostly fixed use to become the main general-purpose federal UHF land mobile bands.

Uses include links to connect mobile networks as well as for weather, hydrologic and seismic sensors, law enforcement and protection of the President and other government personnel and foreign dignitaries, flood and wind-shear warning and telemetry for electric transmission.

Air Force, Army and Navy uses include radar, paging and range control, disaster preparedness, training, shipboard operations, fire protection, transportation and security. The 410–420 MHz portion is exclusive to the government. Radio astronomy is the exclusive non-government user in 406.1–410 MHz portion, one of several bands used for observation of pulsars (see 225–328.6 MHz).

Trends in this spectrum include migration of fixed systems to the 932–935 and 941–944 MHz band, and increased use of multi-agency trunking (automated channel selection) systems. This will result in fewer single-channel, single-user systems. A private corporation operates trunked systems for the government in this spectrum in major cities.

The technology in the 406–420 MHz federal bands will change as the National Telecommunications and Information Administration implements narrowband radios here (see 162.0125–173 MHz).

This is one of many restricted bands in which the FCC's Part 15 rules permit unlicensed devices to emit only very low level emissions.

## 410-420 MHz

Gov: FIXED. MOBILE. Space Research (space-to-space). US13 G5 651A

Non-Gov: US13

See 406.1–410 MHz for uses of this band.

Internationally, 410–420 MHz is available to the Space Research Service on a secondary basis for space-to-space links within five kilometers of an orbiting manned vehicle.

The Space Research Service is defined as "a radiocommunication service in which spacecraft or other objects in space are used for scientific or technological research purposes."

Airborne synthetic aperture radar, for such uses as target identification, operates in 410–450 MHz among its bands.

Person-overboard marine alarms use 418 MHz under FCC Part 15 rules that provide for low-power control devices in several VHF and UHF bands.

## 420-450 MHz

Gov: RADIOLOCATION. 664 668 US7 US87 US217 US228 US230 G2 G8

Non-Gov: Amateur. 664 668 US7 US87 US217 US228 US230 NG135

FCC: LAND MOBILE (90). Amateur (97).

This band is used for telemetry, telecommand and long-range surveillance by land, ship and airborne stations—for early missile warning, detection of low-observable military targets, and tracking of all objects in Earth orbit.

The Air Force Ballistic Missile Early Warning System (BMEWS) radars in Alaska, England and Greenland operate in this band. BMEWS is considered the backbone of the U.S. missile defense system.

Tracking satellites and submarine-launched ballistic missiles, the Precision Acquisition Vehicle Entry Phased Array Warning System (PAVE PAWS) radars operate in this band from Air Force bases in California, Georgia, Massachusetts and Texas.

PAVE PAWS can simultaneously detect and discriminate among many objects and provide data on launch, position, velocity and impact. The band also is used for development of anti-stealth and foliage penetration radars.

The Arecibo Observatory conducts radar and astronomy observations in this band (see 2360–2390 MHz).

The National Oceanic and Atmospheric Administration (NOAA) plans a network of 449 MHz wind profilers across the U.S. They warn of severe wind conditions for forecasts and aviation, and help in analysis of pollutants such as volcanic ash and acid rain. Wind profilers are considered an improvement over radiosondes lofted by weather balloons.

The FCC also is studying how to accommodate non-federal wind profilers at 915 MHz. NOAA and the National Science Foundation operate profilers at this frequency.

The 420–450 MHz band is a popular Amateur Radio Service allocation, especially for FM repeater use in the 440–450 MHz segment.

Amateur use of the band is growing, particularly because of congestion in the 144–148 and 222–225 MHz bands and the increasing availability of dual-band VHF/UHF hand-held amateur transceivers.

The amateur satellite program relies on 435–438 MHz for links with spacecraft. Amateur television is also growing in popularity in this band. We anticipate wider use of spread spectrum technology in the band in the years to come.

Sharing of 421–430 MHz with Private Mobile Radio Services is permitted in Buffalo, Detroit and Cleveland.

## 450-451 MHz

Gov: 668 669 670 US87

Non-Gov: LAND MOBILE. 668 669 670 US87 NG12 NG112 NG124 NG148

FCC: AUXILIARY BROADCASTING (74). SATELLITE COMMUNICATION (25).

This band is used for base stations and 455–456 MHz for mobile units by TV and radio broadcasters for Broadcast Auxiliary operations, wireless microphones and mobile communications in support of electronic news gathering.

Another use is telemetry from broadcast transmitters back to the studios from where the transmitters are controlled.

The footnotes make this band available for uplink and telecommand operations in the space services; however, other bands are more commonly used for these purposes.

## 451-454 MHz

Gov: 668 669 670 US87

Non-Gov: LAND MOBILE. 668 669 670 US87 NG12 NG112 NG124 NG148

FCC: PRIVATE LAND MOBILE (90).

This base station band, with channels for mobile units at 456–459 MHz, marks the start of the principal UHF Private Mobile Radio Services (PMRS) allocations which extend to 470 MHz (and in some areas, 512 MHz).

Modern mobile radio has expanded beyond the 450 MHz UHF bands to the 800 and 900 MHz bands, where trunking and digital technologies are used. There, some former PMRS services have been renamed Commercial Mobile Radio Services. They are competing with cellular and other for-profit mobile radio services.

Nevertheless, the 450–470 MHz UHF spectrum is the heart of conventional mobile radio in the U.S. today. The FCC's refarming initiative will greatly affect this UHF spectrum. Refarming promises to bring improved spectrum efficiency, new technologies, and new entrepreneurial communications businesses in the UHF PMRS bands.

PMRS frequencies in 450–470 MHz are currently spaced 25 kHz apart with secondary, low-power operations allowed on frequencies offset 12.5 kHz from assigned channels. In most of these PMRS frequencies, refarming establishes 6.25 kHz wide channels (see 30.56–32 MHz).

Also operating in 450–470 MHz is biomedical telemetry in hospitals. Patients can remain ambulatory while their vital signs are transmitted to monitoring stations. Use of this spectrum by PMRS causes interference to licensed biomedical telemetry devices.

"Actions taken in the PMRS refarming proceeding threaten to render these channels unusable for biomedical telemetry purposes," manufacturers have stated. The FCC is seeking to better accommodate these devices in the VHF and UHF TV bands.

## *454–455 MHz*

Gov: 668 669 670 US87

Non-Gov: LAND MOBILE. 668 669 670 US87 NG12 NG112 NG124 NG148

FCC: Domestic Public Land Mobile (22). Maritime (80).

This band contains the uplinks used by base stations in the General Aviation Air-Ground Radiotelephone Service, a Public Mobile Service authorized by the FCC in 1969 for business aviation. It is divided into 12 25 kHz channels at 454.700–454.975 MHz. The ground signaling channel is 454.675 MHz. Downlinks are in the 459 MHz band.

The largest operator of 450 MHz air-ground telephones is Mobile Telecommunications Technologies Corp. (Mtel) of Jackson, Miss.

The FCC recognized that the allocations at 450 MHz to this service were too small to provide much capacity for consumer use. Most air-ground communications today is in 849–851 and 894–896 MHz. There, several competitors offer Commercial Aviation Air-Ground services on scheduled airlines.

The 454 MHz band also contains channels for the Public Mobile Rural Radiotelephone Service and the Paging and Radiotelephone Service. The latter includes stations using IMTS technology, the Improved Mobile Telephone Service edged out by cellular. These uses are paired with 459–460 MHz and they also have channels at 152 and 157 MHz.

## 455-456 MHz

Gov: 668 669 670 US87

Non-Gov: LAND MOBILE. 668 669 670 US87 NG12 NG112 NG124 NG148

FCC: AUXILIARY BROADCASTING (74).

See 450–451 MHz for uses of this band. This band and 459–460 MHz were allocated to ITU Region 2 (the Americas) for uplinks in the Non-Voice Non-Geostationary Mobile Satellite Service (see 137–137.025 MHz).

## 456-459 MHz

Gov: 668 669 670 US87

Non-Gov: LAND MOBILE. 668 669 670 US87 NG12 NG112 NG124 NG148

FCC: PRIVATE LAND MOBILE (90).

This mobile band is associated with 451–454 MHz.

## 459–460 MHz

Gov: 668 669 670 US87

Non-Gov: LAND MOBILE. 668 669 670 US87 NG12 NG112 NG124 NG148

FCC: DOMESTIC PUBLIC LAND MOBILE (22). MARITIME (80).

This is the downlink band in the General Aviation Air-Ground Radiotelephone Service (see 454–455 MHz). The band also is used by the Rural Radiotelephone Service and mobile units in the Paging and Radiotelephone Service.

This band and 455–456 MHz were allocated to ITU Region 2 (the Americas) in 1995 for uplinks in the Non-Voice Non-Geostationary Mobile Satellite Service (see 137–137.025 MHz).

## 460–462.5375 MHz

Gov: Meteorological-Satellite (space-to-Earth). 669 670 671 US201 US209 US216

Non-Gov: LAND MOBILE. 669 670 671 US201 US209 US216 NG124

FCC: PRIVATE LAND MOBILE (90).

This band for base stations and the 462.7375–467.5375 MHz band for mobile units encompass much of the UHF spectrum used by businesses and public safety agencies.

Most of the channels in these bands are allocated to the Business Radio Service, one of the most widely used of the Private Mobile Radio Services because of its broad eligibility. Some Business Radio channels in this band are dedicated for airport operations and alarm systems.

The Police and Fire Radio Services have several channels in these bands, and a small number of channels also are allocated to the Forest Products, Petroleum, Telephone Maintenance, Power and Manufacturers Radio Services.

# 462.5375-462.7375 MHz

Gov: Meteorological-Satellite (space-to-Earth). 669 670 671 US201 US209 US216

Non-Gov: LAND MOBILE. 669 670 671 US201 US209 US216 NG124

FCC: PERSONAL (95)

This is one of two UHF segments allocated to the General Mobile Radio Service (GMRS) and the Family Radio Service (FRS) for personal radio communications.

GMRS unit-to-unit and repeater output transmissions use eight primary channels from 462.550 to 462.725 MHz. Repeaters, which extend range over a wide region, use eight input channels from 467.550 to 467.725 MHz.

Overlapping the 462 MHz primary channels are seven low-power offset channels at 462.5625–462.7125 MHz, shared with the Family Radio Service. GMRS also shares 31–31.3 GHz for fixed and mobile use, although the FCC may reallocate that band to a broadband fixed service.

GMRS was created in 1947 as "Class A of the Citizens Radio Service". Class B was a low power version eliminated in 1966. Class C is today the Radio Control Radio Service (see 72–73 MHz).

Class D, created in 1958, was a 27 MHz service later renamed "Citizens Band". The FCC tried and failed to create a 220 MHz Class E service in 1973.

A persistent myth is that GMRS was once "UHF CB" or was "Class A Citizens Band". In fact, the FCC renamed Class A Citizens Radio to the GMRS in 1976, skipping the CB name entirely.

Starting in 1968 with the first system in Chicago, still on 462.675 MHz, the GMRS developed a nationwide mutual-aid network of repeaters, monitors and user groups.

GMRS traditionally used FM radios identical to those used by commercial licensees. Lower-priced, consumer-type GMRS radios from such suppliers as Maxon and Uniden later became available.

The FCC Personal Use Radio Advisory Committee envisioned a 900 MHz personal radio service as an improvement over CB and a possible adjunct or successor to GMRS. That concept became the ill-fated Private Radio Communications Service on which General Electric lost millions (PRCS, see 896–901 MHz).

The FCC later sought to transform GMRS into a so-called Consumer Radio Service including, in the FCC's words, "abolition of repeaters, reduction of maximum authorized power, and the adoption of complex standardized algorithms and protocols." The proposal met with uniform derision among GMRS users and the FCC terminated it.

Eventually the FCC updated the GMRS rules to permit transient (roamer) usage, more channels, and other features recommended by the advisory committee.

Only an individual person may obtain a new GMRS license, or modify an existing one. A licensee's immediate family members are eligible to operate under his or her license. Before these 1989 changes, any legal entity (a corporation, association, a partnership, or a governmental agency) was also eligible to license in GMRS.

In some places, especially in the larger urban areas, these non-personal licensees still aggressively dominate the GMRS channels. The usurpation of the limited GMRS allocation by business users eligible in other services is a continuing problem for GMRS.

The FCC's ban on new non-personal licensing helped to slow the problem somewhat, though the agency does not reliably observe the ban. Officially, the FCC has stated that the GMRS is not "the other Business Radio Service."

Many licensees belong to local cooperatives or teams that provide repeater service at little or no cost. The GMRS National Repeater Guide lists repeaters and their sponsors.

That publication, as well as licensing instructions, a newsletter, and on-line services are available from the Personal Radio Steering Group Inc. (PRSG), PO Box 2851, Ann Arbor MI 48106.

In July 1994, Tandy Corp. (Radio Shack) proposed a Family Radio Service (FRS) which would use spectrum in the GMRS band. The FCC went forward with a Notice of Proposed Rulemaking, its third attempt to establish a new personal radio service in the UHF spectrum.

The prospect of an unlicensed service in the GMRS band shocked GMRS licensees, especially because the FCC proposed few safeguards against abuse of FRS radios. A minority of the formal comments filed on the subject supported FRS. The majority vehemently opposed it.

Prompted by manufacturer interest in radio sales, the FCC claimed there was "overwhelming" support for FRS and established the service on July 8, 1996. FRS operates on 14 offset channels at 462 and 467 MHz.

"The FRS will enable families, friends and associates to communicate among themselves within neighborhoods and while on group outings," the FCC said. FRS "would be useful to hunters, campers, bikers, bicyclists and other outdoor activity enthusiasts who need to communicate with other members of their party who are out of speaking range or sight but still in the same general area."

The PRSG said, "The established GMRS infrastructure, which the FCC already dedicates for personal and family communications, will be at substantial risk of debilitating interference from the unlicensed users—many of them business users—of the Family Radio Service."

## 462.7375–467.5375 MHz

Gov: Meteorological-Satellite (space-to-Earth). 669 670 671 US201 US209 US216

Non-Gov: LAND MOBILE. 669 670 671 US201 US209 US216 NG124

FCC: PRIVATE LAND MOBILE (90).

This band includes mobile units for the business, industrial and public safety systems in 460–462.5375 MHz. Private Carrier Paging Systems are in this band (see 152.255–152.495 MHz).

The band also contains base units in the Emergency Medical Radio Service, MED channels 1 through 10. They convey biotelemetry and voice during patient transport. The mobile MED channels are in the 467.7375–470 MHz band.

## 467.5375–467.7375 MHz

Gov: Meteorological-Satellite (space-to-Earth). 669 670 671 US201 US209 US216

Non-Gov: LAND MOBILE. 669 670 671 US201 US209 US216 NG124

FCC: PERSONAL (95).

This band is part of the General Mobile Radio Service and Family Radio Service (see 462.5375–462.7375 MHz).

## 467.7375–470 MHz

Gov: Meteorological-Satellite (space-to-Earth). 669 670 671 US201 US209 US216

Non-Gov: LAND MOBILE. 669 670 671 US201 US209 US216 NG124

FCC: PRIVATE LAND MOBILE (90).

This band is used by the Business Radio Service and by the Emergency Medical Radio Service for mobile units on MED channels 1–10 (see 462.7375–467.5375 MHz).

The Geostationary Operational Environmental Satellite (GOES) transmits in this band among its downlink frequencies (see 1670–1690 MHz).

## 470-512 MHz

Gov: None

Non-Gov: BROADCASTING. LAND MOBILE. NG66 NG114 NG127 NG128 NG149

FCC: RADIO BROADCAST (TV) (73) PUBLIC MOBILE (22) PRIVATE LAND MOBILE (90) Auxiliary Broadcasting (74)

This band contains UHF television channels 14–20.

Private Mobile Radio Services share this band in 11 cities and the Gulf of Mexico under power and siting restrictions designed to protect TV from interference.

In New York City and surrounding areas TV channels 14 (470–476), 15 (476–482) and 16 (482–488 MHz) will be devoted to public safety and municipal agencies for several years. Eventually a digital TV station will use channel 16. In Los Angeles, the FCC allocated channel 16 outright to public safety radio.

The United States Information Agency broadcasts video propaganda to Cuba over TV Marti. The station transmits from an aerostat 10,000 feet above Cudjoe Key, Florida, on channels 18, 50 and 64. It randomly changes channels in reportedly futile attempts to counter Cuban jamming.

TV Marti has almost never been seen in Cuba. Ever since the station went on the air in 1990, the Cuban government has jammed it. “Federal researchers have searched assiduously and in vain for a TV Marti audience,” according to the *Washington Post.*

The American television system uses the NTSC standard, developed by the National Television Systems Committee for black-and-white TV in 1940–41 and color in 1950–53.

NTSC was nicknamed "Never Twice the Same Color." It prompted generations of viewers to festoon TV antennas with kitchen foil and guaranteed the cable industry a lucrative market in providing adequate reception.

Television will see fundamental transformation over the next two decades. Since 1987, the FCC has studied how to convert the TV system from NTSC to Digital TV (DTV).

High-Definition TV (HDTV), a service available via DTV, will deliver widescreen images at twice the resolution of NTSC, approaching the picture quality of 35mm movies. The HDTV standard also provides for ancillary services such as multiple languages and audio channels, data communications and supervisory and switching information for broadcasters.

A popular myth is that the FCC imposed HDTV on broadcasters. In fact the U.S. HDTV and DTV campaign was an elaborate project of the broadcasting and receiver manufacturing industries.

It began in 1987 with an industry petition to the FCC asking it to investigate advanced TV. At the time, the FCC was seeking to increase the opportunities for sharing of TV spectrum by land mobile radio services.

Such sharing already functions in 470–512 MHz. Known to the mobile industry as the "T-band", it is available in urban areas, where demand is greatest for mobile radio including public safety, industrial and transportation uses.

Broadcasters were alarmed at the prospect of more incursion into TV spectrum by mobile radio. Sharing with the mobile service dilutes, at least in theory, the value of TV licenses and invites more interference. (The FCC eventually conceded publicly that the cause of most TV interference is in the design and construction of TV sets.)

Broadcasters believed that they might stave off further sharing of their spectrum by invoking the exotic specter of HDTV, which could require all of the UHF TV spectrum and might not tolerate more sharing with mobile radio.

HDTV was portrayed as crucial to renewed American dominance in the consumer electronics market. The campaign worked extremely well. Japan and Europe had already launched HDTV technology development with much fanfare, a point not lost on Congress.

The FCC established an official Advisory Committee on Advanced Television Service (ACATS). The broadcast, manufacturing, media and cable industries opened HDTV testing laboratories and established a joint Advanced Television Systems Committee (ATSC).

A Digital HDTV Grand Alliance was formed in 1993 by manufacturers and labs whose competing systems were undergoing independent tests. The organizations are AT&T, General Instrument, MIT, Philips, the Sarnoff laboratories, Thomson and Zenith. They merged their designs into a single DTV system.

The ATSC drafted a standard based on the Grand Alliance system, and in November 1995 the ACATS recommended that the FCC adopt the ATSC DTV standard. In May 1996 the FCC proposed to do so.

The FCC lauded the standard: "Use of the ATSC DTV standard represents a rare opportunity to increase significantly the efficient use of broadcast spectrum. ...

"The packetized structure of the data transport ensures a flexibility that will permit the DTV licensee to provide, for instance, several standard definition programs, or one high-definition program, or some standard definition programming together with data transfer or electronic publishing on the remaining bit streams, and to switch instantaneously between such applications. This means a wide array of innovations can be introduced without Commission action."

The computer industry lobbied heavily to establish progressive scan, an image rendition method, in the DTV standard instead of the interlaced method traditional in television and incorporated into the Grand Alliance design. Computer interests maintained that progressive scan is optimal for integration of computer image and video transmission; TV interests countered that progressive scan is costly and that the Grand Alliance system could accommodate it in the future if necessary.

TV stations will transmit NTSC and DTV simultaneously on separate channels for years. Eventually, the last NTSC transmitter will be turned off. Converters should be available for holdouts unwilling to junk their NTSC sets.

The FCC proposed to consolidate all U.S. DTV in channels 7–51 with some stations operating in channels 52–69 for a limited period. Each existing TV station is to be assigned a channel in this core region. Vacated NTSC channels are expected to be placed on the auction block for uses now undetermined.

TV stations may find it more profitable to broadcast multiple, NTSC-like standard definition programs than to utilize the entire channel for HDTV. Edmund Andrews of the *New York Times* reported, "Broadcasters and regulators increasingly agree that what the American public will want is more television, not better television."

FCC Chairman Reed Hundt described HDTV as "very marginal", "unnecessary", "possibly superfluous" and even "defunct" in view of the non-HDTV uses of DTV channels.

The wireless industry is suspicious that broadcasters will use DTV to offer competitive mobile services. "The fact that broadcasters continue to lust greedily after additional free spectrum to use for any services they choose is certainly misguided, at best, and highway or spectrum robbery, in fact," according to Jay Kitchen, president of the Personal Communications Industry Association.

DTV should offer greater realism and image quality. Not all its features are consumer benefits, however. It can thwart devices intended to "zap" unwanted commercials.

The FCC allows data transmission on the Vertical Blanking Interval (VBI) portion of the TV signal. No picture information is transmitted during the VBI, when the electron beam retraces from the bottom to the top of the screen. The VBI transmits closed captioning and data that broadcasters use internally to track ads and programs.

Other TV data methods are "overscan," which replaces video with data in a part of the picture that is usually invisible to viewers (it is hidden by the frame around the screen); and the "sub-video" method that distributes the data signal at low levels throughout the visible picture.

"Sub-video systems are capable of relatively high data rates, on the order of 300–500 kbps," according to the FCC. "Such high-speed uses include digital magazines, newspapers, books, catalogues, and audio materials, downloading computer software or financial data, and distributing business data to branch offices or clients." One application transmits data from the TV to stuffed animals, causing them to talk.

At a 1993 Senate hearing, a University of Washington psychiatry professor testified that long-term childhood exposure to TV is a causal factor behind approximately one-half of the homicides committed in the U.S.

## 512-608 MHz

Gov: None

Non-Gov: BROADCASTING. NG128 NG149

FCC: RADIO BROADCAST (TV) (73). Auxiliary Broadcasting (74).

This band contains UHF TV channels 21–36.

Biomedical telemetry is allowed on channels 21–29. FCC power limits imposed to protect TV reception preclude use of this allocation, according to telemetry manufacturers.

At their request, the FCC proposed to raise the permitted power for Part 15 biomedical telemetry in the upper VHF TV band and the entire UHF TV band (see 174–216 MHz).

## 608-614 MHz

Gov: RADIO ASTRONOMY. US74 US246

Non-Gov: RADIO ASTRONOMY. US74 US246

No transmissions are permitted on channel 37, a receive-only scientific band.

Radio astronomers use it for Very Long Baseline Interferometry: the connection of telescopes far apart, for observing such sources as pulsars (see 225–328.6 MHz).

The National Radio Astronomy Observatory operates the Very Long Baseline Array of telescopes at ten sites in the U.S. and territories. Data from the antennas are combined, synthesizing a single telescope 5000 miles in diameter.

This is one of many restricted bands in which the FCC's Part 15 rules permit unlicensed devices to emit only very low level emissions.

## 614-806 MHz

Gov: None

Non-Gov: BROADCASTING. NG30 NG43 NG128 NG149

FCC: RADIO BROADCAST (TV) (73) Auxiliary Broadcasting (74)

This band contains UHF television channels 38–69.

The UHF TV band originally reached to channels 70–83 (806–890 MHz). The FCC later reallocated that segment to mobile services and required TV stations in it to move to other channels. Older TV sets that tune channels 70–83 can receive the mobile services such as cellular telephone.

The FCC proposed to eventually reclaim channels 52–69 for other uses as a part of its transition to Digital Television (see 470–512 MHz).

## 806-821 MHz

Gov: 704A US116 US268 G2

Non-Gov: LAND MOBILE. NG30 NG31 NG43 NG63 NG151

FCC: PRIVATE LAND MOBILE (90) PUBLIC MOBILE (22)

This band contains the mobile-to-base frequencies for 800 MHz Specialized Mobile Radio (SMR), the latest generation of private mobile communications. The companion base-to-mobile band is 851–866 MHz.

Together with the developing 900 MHz SMR spectrum (896–901 and 935–940 MHz), these bands contain digital services that challenge cellular and the new Personal Communications Services.

Several services use this spectrum, including the catchall Business Radio Service and the Industrial and Land Transportation and Public Safety services. The bands are mainly recognized for their use by SMR services.

The bands are divided into segments for conventional and "trunked" radio systems. Both types of systems use repeaters, sometimes called "mobile relay stations".

Repeaters receive the signals of mobile and hand-held radios and retransmit them, usually in another band, over a wide area from a high location.

Conventional private land mobile systems typically use one channel (a single frequency, or pair of frequencies in two separate bands). A user needing to transmit must listen and wait until the channel is available to use. This traditional arrangement is inefficient and counterintuitive. A nearby channel may be unoccupied but unavailable for use because it is licensed to other parties.

Trunked systems apply automated selection to groups of channels. The user's radio and the trunked repeater switch to the next available channel without user effort. This technique allows more users to be served per radio channel.

When it decided to release 800 MHz spectrum to private land mobile systems in 1974, the FCC wanted to use trunking on the new channels. Trunked repeater equipment, however, is too complex and expensive to expect individual users to purchase.

To get trunked equipment introduced into the market the FCC established SMR, a new type of business in which licensed operators could sell trunked repeater service and trunked mobile radios to users eligible to hold radio licenses in the private services.

State utility commissioners wanted regulatory authority over SMR and unsuccessfully fought the FCC's creation of it, as did paging and mobile phone companies who had been the only parties allowed to receive subscriber revenue from mobile radio transmissions.

SMR's profit potential provided the needed incentive for entrepreneurs to invest in trunked repeaters and radio sales and service facilities in the new bands. SMR demonstrated the value of trunked radio. The technique was introduced into other services including public safety, cellular and federal land mobile radio.

Other advantages of trunking include selective calling and telephone interconnection. The computerized channel management of a trunked system permits dispatchers to selectively communicate with specific users or groups of users, without disturbing other users or requiring them to listen for calls. This feature is critical to the operation of modern police radio networks, for example.

Trunking also made it easier to offer users reliable telephone interconnection: the ability to make telephone calls from mobile and portable radios.

The FCC ultimately allocated 816–821 MHz for trunked SMR mobile-to-base and 861–866 MHz for trunked SMR base-to-mobile use. It allocated 806–810 MHz and 851–855 MHz for conventional systems. The 810–816 and 855–861 MHz segments were devoted to trunked or conventional systems in the Business, Industrial and Land Transportation, Public Safety, and SMR Services.

Most SMR customers are in construction, service and transportation industries. High-capacity digital technologies are beginning to move SMR beyond dispatch to paging, data, E-mail and cellular-like telephone communications.

This process began when Fleet Call, a large SMR owner founded by former FCC officials, launched what it called the "Enhanced SMR" (ESMR) or wide-area SMR in Los Angeles, using a Time-Division Multiple Access digital architecture. The company changed its name to Nextel to symbolize its transition from dispatching to next-generation mobile telecommunications service. Nextel acquired SMR licenses in the largest U.S. markets.

"ESMR...involves the accumulation of SMR stations, the application of digital technology, and the operation of multiple low-power base stations with significant channel reuse," Nextel said. "These innovations, introduced at a cost of over one billion dollars to Nextel, make possible an advanced mobile communications system capable of providing mobile telephone service comparable to that currently provided by the cellular industry, as well as private network dispatch, paging and mobile data services." Other operators acquired multiple SMR licenses to create wide-area networks similar to Nextel.

Nextel alliteratively dubbed its user device the Fully Integrated Digital Portable Flip Phone (FIDPFP).

The advent of high-capacity, wide-area ESMR services prompted the cellular industry to lobby for laws that would purport to "level the playing field" and subject the two services to the same regulations. This principle is known as "regulatory parity": the similar treatment of like services.

As a result, Congress directed the FCC to similarly treat cellular, ESMR, Personal Communications Services, paging and other telephone-interconnected, subscriber revenue based services on exclusive frequency and market slots. These are Commercial Mobile Radio Services (CMRS). Traditional mobile services for licensees' internal use are termed "Private Mobile Radio Services" (PMRS).

CMRS licenses are auctioned while the FCC grants PMRS licenses on a first-come, first-served, shared-frequency basis. CMRS licensees are now permitted to offer fixed services; in fact, they may legally offer only fixed services and no mobile services at all.

An anticipated fixed CMRS service is wireless local loop, basic phone service to fixed users in competition with local telephone companies. According to the FCC, "Other potential wireless fixed services may include wireless Internet access, electronic funds transfers, point-of-purchase credit card verification, and remote monitoring."

Desiring to promote wide-area ESMR networks using digital technology in lieu of smaller, incumbent FM systems, the FCC designated 10 MHz (816–821 / 861–866 MHz) for SMR systems "that are comparable to and compete with cellular and Broadband PCS systems."

The 10 MHz segment is divided into three blocks of 120, 60 and 20 channels. The FCC will auction the blocks in 175 Economic Areas (EAs), geographic zones defined by the Department of Commerce. Incumbent operators within the EAs are subject to voluntary and, if necessary, involuntary relocation to other SMR spectrum at the expense of the EA auction winner, if the EA licensee can find other available channels in which to relocate them.

In a decision that frustrated many 800 MHz users, the FCC reallocated to SMR a portion of the spectrum known as the General Category (806–809.75 / 851–854.75 MHz), in part to accommodate relocating incumbents.

This segment was available to all eligible 800 MHz users for either conventional or trunked operation, but licenses in it are likely to be auctioned as well. Auction winners will have to accommodate those already using it.

SMR, ESMR and PMRS systems in these bands carry data communications as well as voice. Other services dedicated to mobile data have also emerged. The ARDIS network operates on nationwide and local 800 MHz frequencies, providing communications for field service, transportation, security and industrial customers.

Competing with ARDIS are RAM Mobile Data, a SMR-based network using the Mobitex standard, and Cellular Digital Packet Data (see 824–849 MHz).

Portions of 806–824 MHz are used for high-power Navy shipborne long-range search radars.

## 821–824 MHz

Gov: 704A US116 US268 G2

Non-Gov: LAND MOBILE. NG30 NG31 NG43 NG63 NG151

FCC: PRIVATE LAND MOBILE (90).

To promote technology, interoperability and local control in public safety services such as the Police and Fire Radio Services, the FCC allocated 821–824 (mobile-to-base) and 866–869 MHz (base-to-mobile) to them in a national plan.

The plan establishes a National Public Safety Planning Advisory Committee to oversee regional committees that coordinate frequency use in these bands.

Public safety licensees are implementing digital radio technology, under the "Project 25" standards of an association of public-safety communications officers. This technology provides spectrum efficiency, encryption, selective calling, fast notification of crisis situations, and the ability both to disable stolen radios, and to cause them to surreptitiously transmit as an aid to locating them.

The national plan public safety channels also include channels dedicated to mutual aid, enabling different jurisdictions to readily intercommunicate.

Portions of 806–824 MHz are used for high-power Navy shipborne long-range search radars.

## 824–849 MHz

Gov: 704A US116 US268 G2

Non-Gov: LAND MOBILE. NG30 NG31 NG43 NG63 NG151

FCC: PUBLIC MOBILE (22).

The Cellular Radiotelephone Service, one of the Public Mobile Services (and a Commercial Mobile Radio

Service), provides for two carriers, "A" and "B", in each geographic market—the cellular "duopoly".

Originally the "B" or "wireline" carrier was affiliated with the local telephone company and the "A" or "nonwireline" carrier was not so affiliated. In 1986 the wireline companies were allowed to buy "A" carriers.

The cellular channels begin with 824.04–834.99 MHz, the "A" segment devoted to mobile-to-base transmissions.

In 1970, AT&T proposed a high-capacity cellular phone system based on the reuse of frequencies at multiple base station sites (cells or cell sites). Mobile units communicate with cells, which connect by microwave or landline to a Mobile Telephone Switching Office (MTSO). Together with hardware and software in the cells and mobile units, the MTSO manages signaling, telephone interconnection, channel and power-level selection, and transfer of traffic between base stations.

AT&T called the system AMPS, Advanced Mobile Phone Service, and that term describes the analog cellular standard used in U.S. cellular today. The FCC issued rules for the service in 1981 after years of proceedings. Commercial cellular service in the U.S. began in Chicago on October 13, 1983 when technician Jeff Benuzzi beat 13 others in a race to be the first to complete a customer's car phone installation.

The FCC designated 734 cellular licensing areas, comprised of 306 "Metropolitan Statistical Areas" and 428 "Rural Statistical Areas." By year-end 1995, there were more than 33 million U.S. cellular subscribers, more than 22,000 cell sites, and the cumulative cellular infrastructure investment exceeded $24 billion.

Within the block "A" cellular mobile-to-base segment 824.04–834.99 MHz, control channels that manage the calling process are at 834.39–834.99 MHz.

Cellular base-to-mobile segments are 45 MHz higher in frequency than the mobile-to-base segments. For example, the 824.04–834.99 MHz "A" mobile-to-base segment corresponds to the "A" base-to-mobile segment at 869.04–879.99 MHz. Within that base-to-mobile segment, the control channels are at 879.39–879.99 MHz.

The newer mobile-to-base "A" segment is 845.01–846.48 MHz, corresponding to the newer base-to-mobile "A" segment at 890.01–891.48 MHz.

The original "B" mobile-to-base segment is 835.02–844.98 MHz, with control channels occupying part of the segment at 835.02–835.62 MHz. The corresponding "B" base-to-mobile segment is 880.02–889.98 MHz, with "B" base-to-mobile control channels at 880.02–880.62 MHz.

The newer mobile-to-base "B" segment is 846.51–848.97 MHz, corresponding to the newer base-to-mobile "B" segment at 891.51–893.97 MHz.

To accommodate increasing demand for service, in the mid-1980s the FCC added the newer segments just mentioned. In that proceeding, hotly contested by other claimants to the spectrum, the cellular industry persuaded the FCC that cell-splitting (increased re-use of spectrum) was too expensive a substitute for additional spectrum.

Today the cellular industry widely employs cell-splitting to increase capacity. It also has started to introduce other, spectrum-efficient digital standards. The FCC in 1988 allowed cellular carriers to adopt new technologies as long as they continued to support AMPS users.

The U.S. cellular digital standards include IS-54, a Time-Division Multiple Access (TDMA) approach that divides the spectrum into time slots; and the more recent IS-95 Code-Division Multiple Access (CDMA) standard which distinguishes between users by means of unique codes.

The FCC declined to select any digital cellular standard, so use of these standards will not be uniform across the country. Dual-mode phones are planned that will be able to use AMPS and at least one of the other standards.

Cellular serves wireless data users with Cellular Digital Packet Data (CDPD), an interoperable specification developed by a consortium of cellular carriers. CDPD radios operate through the cellular system to convey data for remote monitoring, credit card verification, database access and Internet applications.

Changes to the cellular rules in 1996 affirmed the privilege of cellular carriers and other Commercial Mobile Radio Service providers to offer fixed telephone service, even without offering mobile telephone service. This policy is expected to smooth the entry of cellular carriers into the local exchange fixed telephone business, although PCS carriers were allocated more spectrum than cellular and so theoretically have greater capacity.

The cellular industry will experience dramatic competitive impacts from the advent of Personal Communications Services (PCS, see 1850–1990 MHz). PCS carriers will compete with cellular especially on price, features and convenience of purchase and use.

Analog transmissions are not permitted in PCS. Therefore, PCS signals are inherently more secure from eavesdropping than are analog AMPS cellular signals, which can be received on unmodified television sets as well as specialized receivers.

---

## *849-851 MHz*

Gov: None

Non-Gov: AERONAUTICAL MOBILE. NG30 NG63

FCC: PUBLIC MOBILE (22).

This is the uplink band of Commercial Aviation Air-Ground Systems, part of the Public Mobile Air-Ground Radiotelephone Service: public phones aboard scheduled airlines. Radios connected to handsets in the cabin link passengers to the phone network through ground stations.

Created by communications entrepreneur and MCI founder Jack Goeken, the Airfone service operated experimentally for years only to have its request for permanent spectrum rebuffed twice by the FCC in favor of other services. Goeken sold Airfone to GTE, which continued to operate Airfone under experimental license.

The Commission later relented and established a permanent spectrum home for Commercial Aviation Air-Ground Systems. It allocated 849–851 (uplink) and 894–896 MHz (downlink) to the service. Of the six available licenses, three are in use: GTE Airfone, In-Flight Phone Corp. (affiliated with MCI), and AT&T Wireless Services.

The service expanded beyond voice telephony, adding data communications, fax, merchandising and news services, video displays, and even ground-to-air calling.

Worldwide, 1670–1675 (uplink) and 1800–1805 MHz (downlink) are allocated for air-ground communications, but the U.S. does not use those bands for that purpose.

## *851-866 MHz*

Gov: 704A US116 US268 G2

Non-Gov: LAND MOBILE. NG30 NG31 NG63

FCC: PRIVATE LAND MOBILE (90)

This band contains the base-to-mobile frequencies for conventional and trunked private land mobile radio services (see 806–821 MHz).

## 866-869 MHz

Gov: 704A US116 US268 G2

Non-Gov: LAND MOBILE. NG30 NG63

FCC: PRIVATE LAND MOBILE (90)

This band contains base-to-mobile frequencies for public safety licensees under the FCC's national public safety radio plan (see 821–824 MHz).

Public safety authorities have recommended that the FCC allocate substantially more spectrum to public safety radio, including an additional national interoperability band for interagency and interjurisdictional communications.

## 869-894 MHz

Gov: 704A US116 US268 G2

Non-Gov: LAND MOBILE. NG30 NG43 NG63 US116 NG151

FCC: PUBLIC MOBILE (22)

Base stations in the Cellular Radiotelephone Service transmit to mobile units in this band. Mobile units transmit 45 MHz lower in frequency (see 824–849 MHz).

Segments in this band include (MHz):

869.04–879.99 "A" base-to-mobile voice channels;

879.39–879.99 "A" base-to-mobile control channels;

890.01–891.48 "A" newer base-to-mobile voice channels;

880.02–889.98 "B" base-to-mobile segment;

880.02–880.62 "B" base-to-mobile control channels;

891.51–893.97 "B" newer base-to-mobile voice channels.

The FCC Part 15 rules permit non-cellular, unlicensed operation in this band at very low power levels. Other Part

15 rules forbid manufacture or importation of scanning receivers that tune 824–849 or 869–894 MHz, to protect privacy of cellular users. Cellular signals can be received on older TV sets (channels 70–83) and other devices.

Footnote US268 provides for Navy shipborne long-range search radars in the 890–902 and 928–942 MHz spectrum.

## *894-896 MHz*

Gov: US116 US268 G2

Non-Gov: AERONAUTICAL MOBILE. US116 US268 NG153

FCC: PUBLIC MOBILE (22).

This is the downlink band for Commercial Aviation Air-Ground Systems, part of the Public Mobile Air-Ground Radiotelephone Service (see 849–851 MHz).

## *896-901 MHz*

Gov: US116 US268 G2

Non-Gov: LAND MOBILE. US116 US268

FCC: PRIVATE LAND MOBILE (90)

This band contains the mobile-to-base frequencies for the 900 MHz version of Specialized Mobile Radio (SMR), established in 1986 to alleviate congestion in the 800 MHz SMR service (see 806–821 MHz). Base stations transmit in 935–940 MHz.

Unlike the 806–821 / 851–866 MHz SMR and private radio spectrum, the 896–901 / 935–940 MHz spectrum was slow to develop.

In 1995 the FCC moved to grant the remaining available licenses in the 900 MHz SMR band by auction. The licenses are based on Major Trading Areas (MTAs), geographic zones defined by Rand McNally.

The auction concluded April 15, 1996, raising more than $204 million for 20 ten-channel licenses in each of 51 MTAs. These licenses will be used to provide fixed and mobile voice and messaging Commercial Mobile Radio Services to business, and possibly consumer, markets.

Incumbent licensees in this spectrum will retain the right to operate under their existing authorizations, but will have to obtain the MTA license or the consent of the MTA licensee to be able to expand their systems.

An unusual service that would have utilized this spectrum was the Citizens Band Land Mobile Private Radio Communications Service (PRCS).

The FCC Personal Use Radio Advisory Committee (PURAC) of the 1970s was charged, among other things, with envisioning a successor to the amok 27 MHz Citizens Band. PURAC recommended a new service at 900 MHz.

The FCC issued a proposal based on the PURAC recommendation. The proposal languished until General Electric asked the FCC to expedite it.

GE proposed PRCS as an FM-based consumer service with advanced features such as telephone interconnection, selective calling and a form of trunking (automated channel selection) that did not require centralized control.

PRCS was slated to use 133 channels at 898–902 and 937–941 MHz. GE invested heavily to develop PRCS mobile and base radios. GE demonstrated the system—which it called CarFone—in Washington, D.C.

PRCS was to employ base stations connecting to the public network at owners' premises. It would have been capable not only of phone calls through these stations and via optional repeaters, but also of direct, free mobile-to-mobile and base-to-mobile communication. This feature is impossible to deliver in the public cellular service.

General Mobile Radio Service licensees (see 462.5375–462.7375 MHz) supported PRCS. Cellular carriers and GE's competitors lobbied bitterly against PRCS, however. GE itself delivered a serious blow in 1984 by closing its PRCS project after financial reviews within the company.

Without a manufacturer to build the radios, and with cellular interests determined to halt PRCS, the FCC rejected PRCS in a tense public meeting.

At the last minute some Commissioners seemed reluctant to cancel PRCS. This prompted an outburst from a former FCC official in the audience. He reviled the proposal, warning that if the public were allowed access to PRCS, the FCC would never be able to control it.

## *901-902 MHz*

Gov: US116 US268 G2

Non-Gov: FIXED. MOBILE. US116 US268 US330

FCC: PERSONAL COMMUNICATIONS (24)

The General Purpose Mobile Radio Service (GPMRS) was an ephemeral service at 901–902 and 940–941 MHz with no licensees, rules or operations.

The FCC believed it could create a new type of service in this band open to "all mobile uses". It defended the GPMRS against petitions for reconsideration from land mobile user groups that wanted the GPMRS spectrum.

The FCC did nothing to actually implement GPMRS, however. In 1993 the FCC eliminated the service entirely.

The FCC reallocated the GPMRS bands, plus 930–931 MHz, a former reserve band for paging, to the Narrowband Personal Communications Services (PCS).

The FCC based the decision on a proposal of Mobile Telecommunication Technologies Corp. (Mtel) for a network for portable two-way data devices.

Mtel won a hotly-contested Pioneer's Preference—including early assurance of a free license—from the FCC for its development of multicarrier modulation and other technologies. The FCC said, "Mtel's development of technology permitting attainment of 24 kbps data rates for a nationwide simulcast paging and messaging integrated service is a significant communications innovation... Mtel has demonstrated the feasibility of its new spectrum-efficient technology and designed a system based upon it that permits using the same infrastructure to provide both substantially improved and more spectrum-efficient paging services and new one-and two-way messaging services."

The Mtel network is operated by subsidiary SkyTel. The company is deploying thousands of base stations—considerably more receivers than transmitters.

The FCC organized the Narrowband PCS spectrum into channel sets of different kinds with licenses that it auctioned. For nationwide operation the FCC allocated three 50 kHz channels paired with 12.5 kHz channels, five 50 kHz channels paired with 50 kHz channels (Mtel obtained one of these) and three 50 kHz unpaired channels. It provided 13 channels for regional licenses and ten for local licenses, of which eight are for use by existing paging licensees to add to their systems.

In its first-ever license auction in July 1994, the FCC received $617 million for the ten available nationwide licenses. Winning bidders included Mtel, AT&T Wireless, AirTouch Paging, MobileComm, PageMart and PageNet. The narrowband PCS business is in its early stages, with SkyTel in operation and the other licensees still developing their networks and services.

Mtel was supposed to receive a free Pioneer's Preference license, but the FCC charged the company tens of millions of dollars for it anyway after receiving authority to auction licenses. A federal appeals judge said the FCC's action was "lawless" and "borders on the outrageous".

Footnote US330 limits fixed services in the indicated bands to ancillary services only. The FCC is expected to remove this limitation as it deregulates the Commercial Mobile Radio Services.

## 902-928 MHz

Gov: RADIOLOCATION. 707 US215 US218 US267 US275 G11 G59

Non-Gov: 707 US215 US218 US267 US275

FCC: Private Land Mobile (90) Amateur (97)

The 902–928 MHz "kitchen sink" band accommodates, not always harmoniously, many applications from wireless stereo speakers to military radars, to commercial Location and Monitoring Service (LMS) systems.

This is considered an Industrial, Scientific and Medical (ISM) band (the FCC describes the ISM frequency as 915 ± 13 MHz). ISM equipment includes industrial heaters that cure adhesives, inks and rubber; welding equipment; food equipment such as bacon dryers and donut fryers; and medical instruments such as magnetic resonance imagers and diathermy (tissue-heating) instruments for physical therapy and cancer treatment.

Older models of consumer microwave ovens operate at 915 MHz. Most microwave ovens in the U.S. operate in the 2400 MHz ISM band. (The bands allocated for ISM use in the U.S. are 6.765–6.795 MHz, 13.553–13.567 MHz, 26.957–27.283 MHz, 40.66–40.70 MHz, 902–928 MHz, 2400–2500 MHz, 5.725–5.875 GHz, 24–24.25 GHz, 61–61.5 GHz, 122–123 GHz and 244–246 GHz.)

Military radar is primary in 902–928 MHz, subject to interference from ISM. The military systems include air surveillance radars used on aircraft carriers, other ships and shore installations; perimeter protection devices; data and video links and systems that track and control missiles and vehicles at test ranges.

NASA uses the band with remotely-piloted aircraft that measure atmospheric ozone at altitudes up to 85,000 feet. Other telemetering in this band supports remote sensing imagery for forest fire studies and disaster assessments.

Federal law enforcement agencies make extensive use of 902–928 MHz for video surveillance and related uses such as viewing of the image seen by a bomb disposal robot.

Following 902–928 MHz military radiolocation uses in priority are non-military government radiolocation uses, government fixed and mobile uses, LMS, Amateur Radio, and unlicensed devices, in that order.

The FCC defines LMS as "the use of non-voice signaling methods to locate or monitor mobile radio units. LMS systems may transmit and receive voice and non-voice status and instructional information related to such units."

A "multilateration" LMS system is "designed to locate vehicles or other objects by measuring the difference of time of arrival, or difference in phase, of signals transmitted from a unit to a number of fixed points or from a number of fixed points to the unit to be located."

Multilateration LMS for vehicle location (and potentially, location of people and objects) will expand in this band as new entrants build their systems. Encouraging this trend are the boom in mobile data communications, the improved cost/performance of modern spread spectrum technology and the FCC update of early rules from the 1970s. This update, still under appeal at this writing, seeks to clarify the technical and licensing rules in this service.

The FCC made LMS the first (and so far, the only) service in the Transportation Infrastructure Radio Services. TIRS is supposed to accommodate Intelligent Vehicle Highway Systems, now called Intelligent Transportation Systems, for safety, energy efficiency and navigation. "As we allocate additional spectrum or create new services intended to further the efficiency of the nation's transportation infrastructure, these new services will likely be regulated under the TIRS," the FCC said.

The best known LMS licensee is Teletrac, now providing fleet tracking and stolen vehicle location to subscribers in several U.S. cities. Fleet tracking allows dispatchers of vehicles such as taxis, trucks, buses and ambulances to locate the vehicles and to be notified of emergencies.

A vehicle theft alarm will alert Teletrac if not deactivated within a short time after it is triggered. Teletrac then attempts to contact the vehicle owner and police.

Although it provides a similar service, LMS at 902–928 MHz should not be confused with the Stolen Vehicle Recovery Service (SVRS), operated only by Lo-Jack Corp. SVRS is allocated the single VHF frequency 173.075 MHz, formerly used for federal undercover law enforcement.

Another interesting form of LMS in 902–928 MHz is the Newchild system, used to locate lost children in theme parks and zoos. The missing person wears a small transponder that is queried by fixed transmitter sites (as many as 200) within the park.

A "non-multilateration LMS system" is a system that "employs any of a number of non-multilateration technologies to transmit information to and/or from vehicular units," according to the FCC's ponderous definition. In practice this usually means short-range data communication with radio tags. Readers illuminate the tags affixed to vehicles or other objects.

Passive tags reflect the signal back to the reader's receiver. The tag uses data about the host object in its memory to modulate the reflection ("modulated backscatter"). Some tags use battery powered electronics. The reader recognizes tags by identification of the signals, and forwards this data to computers for processing.

Tag suppliers claim annual sales of $500 million, for uses in manufacturing, hospitals, antipilferage, security and access control, shipping and even identifying lost pets.

For that application, the industry is developing standards for tags implanted in animals. The tags operate under LMS or under Part 15 rules, especially at 2400–2483.5 MHz.

This band has not yet become a major hub of amateur operation as have lower VHF and UHF bands common in amateur mobile and data communications. The first authorized U.S. amateur transmissions in 902–928 MHz were made by equipment imported by this writer.

Low-power unlicensed devices operating under Part 15 of the FCC Rules in 902–928 MHz have become a sizable industry. Applications include residential and business cordless telephone systems; wireless local-area and metropolitan-area networks for computers; fire and security alarms and theft detection systems; wireless bar-code readers for retail, storage and shipping; data collection terminals for industrial, medical and financial uses; remote utility meter readers; remote video cameras and wireless video products for consumers.

Probably the most bizarre use of this band was in spread spectrum remote control of robot jockeys in indoor horse racing. The robots, essentially boxes with antennae and arm-like appendages, even wore jockey hats.

The FCC permits devices using spread spectrum technology to transmit up to one watt of power in the ISM bands. Spread spectrum emissions are produced by frequency hopping or combining data streams with fast digital codes or combinations of these techniques.

As the name implies, spread spectrum emissions are wider in bandwidth than conventional emissions. This makes the spread spectrum system a lessened interference threat to other devices or services and more immune to interference from them.

Unlicensed devices operate on a secondary basis and must accept any interference they may receive from transmitters higher in priority. Nevertheless, manufacturers of 902–928 MHz Part 15 devices are extremely concerned about interference from LMS and the effects of their devices on LMS networks.

In updating its earlier LMS rules, the FCC established a complex regulatory scheme to govern the relationships between licensed and unlicensed emitters in this band. An unprecedented "safe harbor" policy deems spread spectrum Part 15 devices in 902–928 MHz to not interfere with LMS if the devices meet antenna and power restrictions.

The decision was not universally praised. FCC Commissioner James Quello described it as "but our first step into a potential bog of interference problems."

Some manufacturers are shifting to the 2400–2483.5 MHz and 5.725–5.875 GHz ISM bands. These bands also are shared with interferers that threaten long-term viability, by some accounts.

As an ostensibly hospitable alternative, in 1994 the FCC allocated a dedicated band to unlicensed voice and data Personal Communications Services devices at 1910–1930 MHz. That band is supposed to be eventually cleared of fixed microwave stations presently operating there.

Migration of the fixed stations will ultimately enable the band to be exclusively devoted to low-power unlicensed use by cordless telephone and computer products.

However, the uncertainties, huge costs and delays of this band-clearing process prompted computer makers to petition for, and receive, an additional band for unlicensed wireless data at 2390–2400 MHz. The primary service in that band is Amateur Radio.

The 902–928 MHz, 2400 MHz and 5.7 GHz ISM bands are commonly believed to be the only bands available for unlicensed use. Part 15 actually permits very low-power operation in bands through much of the spectrum except for certain restricted bands (identified in SpectrumGuide) and used by safety, science or government services.

Most manufacturers concentrate instead on higher-power spread spectrum products in the ISM bands. Some products on the market in 902–928 MHz do not use spread spectrum, such as some types of cordless telephones.

The FCC proposed in February 1996 to make several modifications to the Part 15 spread spectrum rules. These include changes in the operating parameters of 902–928 MHz devices to better cope with LMS interference; and eliminating a limit on the use of directional gain antennas in commercial and industrial fixed point-to-point systems at 5.7 GHz. The agency also proposed strengthening its rules forbidding the use of external power amplifiers or antennas not a part of the original, FCC-certified Part 15 system. Apparently, the sale of illegal amplifiers and antennas for use with Part 15 devices is increasing.

The FCC is studying whether 902–928 MHz can accommodate some types of wind profiler radars used for weather forecasting, aviation warning and environmental studies. These radars also operate at 449 MHz.

Radio amateurs and the Part 15 community are concerned about the impact of these radars on their operations. The National Telecommunications and Information Administration has stated, "The expansion of use in the 902–928 MHz band by Federal and non-Federal users, including the operation of wind profiling radars, may make this band untenable for amateur operations in the future."

"The 902–928 MHz band is already an uncomfortable melting pot of uses," according to the American Radio Relay League. "The allocation status of this band was not the result of any apparent plan or design.

"It was, rather, the result of incremental allocation decisions, each one addressed to a particular effort to accommodate individual uses. Each additional incremental addition to the 902–928 MHz allocation, now including wind profiler radars and LMS, brings the band closer to 'gridlock' status."

## 928–929 MHz

Gov: US116 US215 US268 G2

Non-Gov: FIXED. US116 US215 US268 NG120

FCC: PRIVATE LAND MOBILE (90) PRIVATE OPERATIONAL FIXED MICROWAVE (94) PUBLIC MOBILE (22)

Since the 1980s, this band, along with the 952–953 MHz band and other nearby bands, has been used by Multiple Address Systems (MAS) for communication between a central point and surrounding points.

Typically 928–929 MHz is used for the directional, remote stations and 952–953 MHz is used for the omnidirectional master station. Service may be self-provided or provided by a MAS carrier to a user. Channel bandwidths are typically 12.5 kHz, although 25 and 50 kHz may also be authorized and the channels may be subdivided to create additional frequencies.

Many MAS uses are described as System Control And Data Acquisition. The utility and petroleum industries use MAS to monitor lines for faults or blockages and for remote meter reading of energy usage. Airports use MAS to control runway lights. Security firms use MAS for alarms.

Banks use MAS for fund transfers from automatic teller machines, gaming agencies use MAS for lottery terminals, and retailers use MAS for credit card verification.

The FCC does not permit MAS to be used to distribute video entertainment, but the service is used to transmit background music for users such as restaurants and offices.

Channels in the 928–929, 941–944 and 952–953 MHz bands are allocated to shared "pools" for use by public, private and government systems of various types. For example, Public Mobile providers may use these channels to remotely control paging or radiotelephone base stations.

Footnote US268 provides for Navy shipborne long-range search radars in the 890–902 and 928–942 MHz spectrum.

## *929-930 MHz*

Gov: US116 US215 US268 G2

Non-Gov: LAND MOBILE. US116 US215 US268 NG120

FCC: PRIVATE LAND MOBILE (90) PUBLIC MOBILE (22)

This is one of the bands of Private Carrier Paging Systems (PCPS). They offer paging and messaging as do Public Mobile (radio common carrier) paging systems.

Both types of paging are under a common regulatory umbrella: Commercial Mobile Radio Services (CMRS). However, the FCC still keeps PCPS and Public Mobile as separate categories.

The FCC proposed to convert PCPS and Public Mobile paging from station-by-station licensing to geographic-area licensing, as used in other forms of CMRS such as cellular and Personal Communications Services. Geographic-area licensees have considerable flexibility to locate, build and modify their systems.

"We believe that geographic licensing is particularly suitable for paging," the FCC said, "because the service has evolved away from single-site systems toward multi-site systems that cover large geographic areas. ...

"Geographic licensing also would (1) enhance regulatory symmetry between paging and narrowband PCS and (2) eliminate inefficiencies in our current rules that effectively create a vast web of relationships between applications for individual transmitter sites at various locations."

The FCC proposed that incumbent systems, operating under station-based licenses, would continue to operate. Geographic licensees, which would win their licenses by auction, would negotiate with incumbents for purchase or relocation of the incumbent stations.

## 930-931 MHz

Gov: US116 US215 US268 G2

Non-Gov: FIXED. MOBILE. US116 US215 US268 US330 NG120

FCC: PERSONAL COMMUNICATIONS (24)

This is one of three bands allocated to the Narrowband Personal Communications Services (PCS), for advanced paging and messaging applications (see 901–902 MHz).

The strange Space Shot service once tried to get access to this band. Space Shot called itself "The Interplanetary Communication System That Links You To The Universe."

Space Shot helped customers "search for extraterrestrial life" by sending love notes or peace messages into space on 903.0125 MHz for $5.00 per "launch." Its purveyors held a temporary, experimental license. The FCC declined its request for permanent status at 930 MHz in 1985.

The Commission dryly said it doubted that the Space Shot transmitter could activate receivers at the target locations, which were supposed to be the Sun, the Moon and "any of the planets in our Solar System." We considered Space Shot an intriguing test of the FCC's pro-business philosophy, but industry commenters called it "a radio version of the Pet Rock."

A Vermont company has resurrected this concept. Its Prayers Heavenbound service transmits religious messages into space for a fee.

"Our exact communications frequency to God's Heaven is the result of much research and is therefore proprietary," the company told us. "Our prayer uplinking facility is not located on United States soil, or within any of its possessions. As such, its existence or RF emissions are of no jurisdictional concern to the FCC."

## 931-932 MHz

Gov: US116 US215 US268 G2

Non-Gov: LAND MOBILE. US116 US215 US268 NG120

FCC: PRIVATE LAND MOBILE (90) PUBLIC MOBILE (22)

This band contains frequencies for internal paging within organizations, and for Commercial Mobile Radio Service paging in the Private and Public Mobile Services. These include frequencies assigned to the original, non-auctioned nationwide paging licenses, currently held by Mtel, BellSouth and ProNet.

One of the most innovative of these services is Electronic Mail Broadcast to A Roaming Computer (EMBARC). Motorola purchased the stalled Megamessage network on 931.9125 MHz and developed EMBARC, a subsidiary company, to provide text services on that frequency.

Motorola later sold the license to ProNet, a paging carrier that also tracks criminals and stolen goods (see 216–220 MHz). Motorola said it intends to continue EMBARC as a content provider, with ProNet among its resellers.

## 932-935 MHz

Gov: FIXED US215 US268 G2

Non-Gov: FIXED US215 US268 NG120

FCC: DOMESTIC PUBLIC FIXED (21) PUBLIC MOBILE (22)

Multiple Address Systems operate in 932–932.5 and 941–941.5 MHz, providing communications for such uses as credit card verification, alarms and remote control.

The common carrier Point-to-Point Microwave Service and the Private Operational Fixed Service are both permitted at 932.5–935 and 941.5–944 MHz. Many of these non-government uses support paging systems.

Government agencies use the band in land mobile radio systems and telemetry. The largest users are the Federal Aviation Administration, the Agriculture Department, and the Department of Energy which uses the band in control of electric power and weather data transfer.

## 935-940 MHz

Gov: US116 US215 US268 G2

Non-Gov: LAND MOBILE US116 US215 US268 NG120

FCC: PRIVATE LAND MOBILE (90) PUBLIC MOBILE (22)

See 896–901 MHz for information about this SMR band.

## 940-941 MHz

Gov: US116 US268 G2

Non-Gov: FIXED. MOBILE. US116 US268 US330 NG120

FCC: PERSONAL COMMUNICATIONS (24)

This is one of three bands allocated to the Narrowband Personal Communications Services for advanced paging and messaging applications (see 901–902 MHz).

## 941-944 MHz

Gov: FIXED. US268 US301 US302 G2

Non-Gov: FIXED. US268 US301 US302 NG120

The 941–944 MHz spectrum is shared for fixed services (see 932–935 MHz). Some older audio Broadcast Auxiliary stations use 941.5–944 MHz (see 944–960 MHz).

## 944-960 MHz

Gov: None

Non-Gov: FIXED. NG120

FCC: AUXILIARY BROADCASTING (74) PUBLIC MOBILE (22) INTERNATIONAL PUBLIC FIXED (23). PRIVATE OPERATIONAL FIXED (94)

Radio and TV stations use 944–952 MHz for such Broadcast Auxiliary uses as studio-transmitter and intercity relay links. They can lease bandwidth for other purposes.

Most of the growth in this segment comes from FM stations switching to wireless links from dedicated telephone lines that connect studios to transmitters.

The 952–960 MHz segment mainly is used by master stations in Multiple Address Systems (see 928–929 MHz) and by private microwave systems in industrial use.

## 960-1215 MHz

Gov: AERONAUTICAL RADIONAVIGATION. 709 US224

Non-Gov: AERONAUTICAL RADIONAVIGATION. 709 US224

FCC: AVIATION (87).

This band is used for commercial, space and military flight operations and air traffic control, including tracking, navigation, collision avoidance and landing guidance.

Air Traffic Control Radar Beacon Systems (ATCRBS) operate on 1030 and 1090 MHz. ATCRBS uses a ground-based transmitter to interrogate airborne transponders and display target information to controllers.

ATCRBS will be replaced by Mode S, also at 1030 and 1090 MHz. This system more selectively interrogates aircraft within its field of view, and provides data communication between aircraft and ground controllers.

Mode S will ultimately provide a nationwide, two-way data capability for air traffic, flight information and surveillance, according to the FAA. A related technology in this band is IFF, Identification Friend or Foe.

The Traffic Alert and Collision Avoidance System (TCAS) uses these same frequencies. This airborne system obtains data about nearby aircraft from transponders. Various types of TCAS equipment can alert pilots to the presence of other aircraft and can recommend maneuvers.

A key military use of this band is the Joint Tactical Information Distribution System. This multibillion-dollar development is part of a NATO system of secure and jam-resistant frequency-hopping communications, navigation and identification among airborne and surface-based elements in hostile environments.

The 960–1240 MHz spectrum is one of many restricted bands in which the FCC's Part 15 rules permit unlicensed devices to emit only very low level emissions.

## 1215-1240 MHz

Gov: RADIOLOCATION. RADIONAVIGATION-SATELLITE (space-to-Earth). 713 G56

Non-Gov: 713

The 1215–1400 MHz spectrum is used by the Joint Surveillance System, a long-range Air Route Surveillance

Radar network operated by the Air Force and the Federal Aviation Administration for air defense and air traffic control. This also is a key band for other radars of the Air Force and Marine Corps.

The ARSR-4 is an advanced new radar in this spectrum being installed around the U.S. by the Air Force and FAA. It can detect small, low-flying aircraft and missiles at long distances and provide height, bearing and distance data.

Through digital signal processing and exotic antenna technology, the ARSR-4 can discriminate between targets and background clutter, especially turbulent water.

The Cobra Dane phased-array radar on the far end of the Aleutian Island chain uses this band to observe test launches of ballistic missiles, and to provide warning and assessment of missile attacks on the U.S. and Canada. The pyramid-like radar can track up to 300 incoming warheads and up to 200 satellites simultaneously.

NASA uses the band for synthetic aperture radar and for satellite remote sensing of ocean surface characteristics.

The frequency 1227.6 MHz (Link 2 or "L2") is one of three downlink frequencies associated with the Navstar Global Positioning System (GPS). The others are L1, 1575.42 MHz, and L3, 1381.05 MHz. L1 and L2 transmit positioning information. L3 broadcasts nuclear explosion alerts (see 1350–1400 MHz).

GPS is the Air Force's $10 billion constellation of 24 satellites to provide latitude, longitude, altitude, velocity and time to land, sea and air users. Congress mandated that all military land vehicles, vessels and aircraft incorporate GPS receivers. Even bombs may be retrofitted with GPS.

Though created for defense GPS launched a successful industry in civilian vehicle location, camping and exploration receivers, navigation, surveying and mapping, timekeeping, software and services. According to the

RAND Corporation, by the end of the 1990s the market for civilian GPS applications is expected to be eight times the size of the market for military applications.

Controlled from Falcon Air Force Base, Colorado Springs, Colorado, each satellite continuously transmits spread spectrum signals on L1 and L2. The user's GPS receiver measures the time of arrival of signals from the satellites and reports the user's position, velocity and time.

The receiver displays position information in the form of coordinates on the Earth or, in some models, alternate indicators such as a map or the world grid system used by radio amateurs. The user location may be recorded or transmitted to others on terrestrial frequencies.

GPS provides two position services: the Standard Positioning Service (SPS) and the Precise Positioning Service (PPS). The SPS, using frequency L1, is the standard level of accuracy available worldwide without restriction to any user.

The GPS command authority can intentionally degrade SPS to deprive adversaries of high accuracy. Several levels of degradation and techniques (such as encryption and "dithering" or skewing the timing signals) are available.

This degradation is called Selective Availability (S/A). Without S/A, SPS accuracy is 30 meters. With S/A activated, SPS accuracy is 100 meters.

The National Security Council announced in March 1996 that "it is our intention to discontinue the use of GPS Selective Availability within a decade in a manner that allows adequate time and resources for our military forces to prepare fully for operations without S/A."

Abandoning S/A will offer civil users higher accuracy. It requires the U.S. to develop other means, such as jamming of GPS, for use against hostile forces. Tests have shown that properly placed jammers can cause an unprotected GPS-aided weapon to completely miss its target.

The PPS uses both L1 and L2 and provides accuracy of 21 meters. PPS is for defense and authorized civil users.

Some types of GPS errors are monitored internally within the GPS satellites and reported via a "health message" for SPS and PPS users. Other failures are detectable only by GPS controllers. It can take from 15 minutes to several hours before users are notified of a problem.

Reliance on erroneous GPS signals could be dangerous for some uses such as harbor navigation. Users can overcome this problem and obtain high accuracy by using differential GPS (DGPS). This technique is based on the location of a surveyed reference station. Information based on the difference between this reference location and the GPS-computed location is transmitted over non-GPS media to DGPS receivers, which apply the corrections.

DGPS permits use of GPS satellites that are not fully functioning. It offers accuracy even greater than the PPS. It can also indirectly circumvent the ability of the United States to degrade the GPS signal.

The Coast Guard broadcasts differential corrections in a standard data format in U.S. coastal areas, the Great Lakes, Alaska and Hawaii on 285–325 kHz radiobeacons. Commercial vendors broadcast the corrections on FM radio subcarriers in 88–108 MHz.

The Transportation and Defense departments are considering a possible fourth GPS signal, for civilian use, in the next generation of GPS satellites. Future receivers could compare the signals and correct errors caused by signal propagation through the atmosphere.

A gradual transition to GPS-based air navigation incorporates WAAS, the Federal Aviation Administration's

$500 million Wide Area Augmentation System. WAAS will operate 24 ground reference stations around the U.S.

These stations compute their GPS-derived locations and compare them with the surveyed locations. A master station receives this data and sends corrections to aircraft via Inmarsat satellites, over the L1 frequency used by GPS.

The system will permit accuracy to within ten meters. Full WAAS operation is expected by 2000. An even more precise Local Area Augmentation System will use psuedosats—GPS-like ground transmitters—in areas such as airports where higher accuracy is needed.

GPS also will aid the transition to Free Flight, a future mode that will reduce direction by ground controllers. Pilots will have more flexibility to choose routes. They will be able to view aircraft separation data in the cockpit from reports of aircraft positions computed with GPS.

Free Flight depends on new communications systems, known generically as Data Link. They incorporate Automatic Dependent Surveillance, which uses GPS and transmits aircraft position to other aircraft and controllers.

The FAA testified before Congress that the government will save more than $200 million annually when today's ground-based navigational aids are replaced by GPS.

GPS received much exposure when reports claimed that adversaries could use it to direct weapons at U.S. targets. Even the accuracy of GPS in its worst-case S/A mode would be sufficient for direct hits on civilian targets and population centers. Military analysts believe that differential GPS would not be needed for this purpose.

The 960–1240 MHz spectrum is one of many restricted bands in which the FCC's Part 15 rules permit unlicensed devices to emit only very low level emissions.

# 1240-1300 MHz

Gov: RADIOLOCATION. 664 713 714 G56

Non-Gov: Amateur. 664 713 714

FCC: Amateur (97).

This band is part of the 1215–1400 MHz spectrum used by Air Force and Federal Aviation Administration radars. Many of these are frequency-agile and operate at high peak power levels (tens or hundreds of kilowatts for defense radars, and one to five megawatts for FAA radars). Their emissions can extend more than 10 MHz in bandwidth.

Other uses include Amateur Radio, synthetic aperture radar and satellite remote sensing of the ocean surface.

A medical device, the Prostatron, uses 1296 MHz for thermotherapy to treat prostate disease. This non-surgical technique delivers microwave energy to selectively destroy prostate cells without damaging other surrounding organs.

The inventors said they chose this frequency "specifically because its wavelength is uniquely suited to the treatment process." Other frequencies in the 915 MHz and 2450 MHz industrial, scientific and medical (ISM) bands resulted in too little and too much energy absorption by the body.

The Prostatron's manufacturer sought a waiver from the FCC to permit its use in the U.S. after showing that the likelihood of interference to other band users is low.

As the 23 centimeter ham band, 1240–1300 MHz is used in mobile, video, digital and experimental communications. These include satellite uplinks in 1260–1270 MHz.

AMSAT Phase 3-D will use this and other amateur bands. The most ambitious amateur space system yet, the $4 million satellite is funded and constructed by the Washington, D.C.-based Radio Amateur Satellite Corp. and its affiliates in several countries.

## 1300-1350 MHz

Gov: AERONAUTICAL RADIONAVIGATION. Radiolocation. 717 718 G2

Non-Gov: AERONAUTICAL RADIONAVIGATION. 717 718

FCC: AVIATION (87).

This is part of the 1215–1400 MHz radar spectrum (see 1215–1240 MHz). Radio astronomers use 1330–1350 MHz to detect hydrogen in galaxies.

The 1300–1427 MHz spectrum is one of many restricted bands in which the FCC's Part 15 rules permit unlicensed devices to emit only very low level emissions.

## 1350-1400 MHz

Gov: FIXED. MOBILE. RADIOLOCATION. 714 718 720 G2 G27 G114 US311

Non-Gov: 714 718 720 US311

The Air Force is the largest government user of 1350–1400 MHz, for high-power, long-range air surveillance radars. The Federal Aviation Administration and the Defense Department operate the Joint Surveillance System radar network in this band around the U.S. perimeter for radionavigation, air and fleet defense and drug interdiction.

Army and Navy uses of this band include headquarters connectivity, communications training, air traffic control, antenna testing, atmospheric research and tactical radio relay. Remote sensing satellites use 1370–1427 MHz to measure soil moisture and ocean salinity.

The 1390–1400 MHz segment will be reallocated from the government to the private sector in 1999 for uses yet to be determined. Federal radar systems at 17 sites will continue in the segment for 14 additional years.

Only uses that do not interfere with remaining operations or radio astronomy will be allowed. This requirement constrains commercial development of 1390–1400 MHz. According to the FCC, it may be difficult to provide commercial service within hundreds of kilometers of the 17 sites. Some sites include or are near major urban markets.

A little-known function of the Global Positioning System (GPS, see 1215–1240 MHz) uses "L3", 1381.05 MHz, to transmit alerts as part of the $1.5 billion Nuclear Detonation Detection System (NUDET).

NUDET transmits an alert by ground command or upon detection of a nuclear event. To minimize interference to radio astronomy, satellite transmissions on this frequency are limited to actual nuclear bursts, tests and training.

The global coverage and reliability of GPS make it an excellent platform for detection and relay of nuclear data. GPS satellites carry X-ray sensors, electromagnetic pulse detectors and the appropriately named "bhangmeters," sensors that detect bursts of light from nuclear explosions.

Other defense satellites may also convey data to NUDET. Augmenting NUDET will be a new program of space-based nuclear explosion sensors, the Global Verification and Location System, on frequencies in this band.

Atmospheric testing of nuclear devices has been banned by international agreement for decades. The capability to monitor clandestine tests as well as aggressive launches is still necessary. The U.S. employs a vast network of remote sensing satellites, domestic and overseas ground-based radars, underwater sound detectors on submarines and eavesdropping stations to detect nuclear events.

Air Force data links in 1350–1400 and 1427–1435 MHz rebroadcast aircraft position during tests and training. This system will be deployed at all major military aircraft and missile test centers. It is used in flight testing of aircraft such as the B-1 and B-2.

This band also is important to radio astronomers. Shifts in the 1420.4 MHz hydrogen spectral line caused by motion of celestial bodies places most of the energy from distant galaxies into 1350–1400 MHz.

The 1300–1427 MHz spectrum is one of many restricted bands in which the FCC's Part 15 rules permit unlicensed devices to emit only very low level emissions.

## 1400-1427 MHz

Gov: EARTH EXPLORATION-SATELLITE (Passive). RADIO ASTRONOMY. SPACE RESEARCH (Passive). 722 US74 US246

Non-Gov: EARTH EXPLORATION-SATELLITE (Passive). RADIO ASTRONOMY. SPACE RESEARCH (Passive). 722 US74 US246

No transmissions are permitted in this receive-only scientific band. In addition to its use by remote sensing satellites, this band contains the hydrogen spectral line at 1420.4 MHz. Astronomers observe this frequency to study distribution and motion of mass in the universe.

Certain survey programs in the Search for Extraterrestrial Intelligence (SETI) concentrate on 1420–1660 MHz. "Thousands of us may have to work for years to hear The Signal, if we ever do. SETI is definitely a gamble, with long odds but high stakes," according to the SETI League, which coordinates monitors who listen for possible signals from advanced civilizations. SETI is recognized in footnote 722, which applies to this and adjacent bands.

Out-of-band and spurious emissions from 1350–1400 MHz, especially commercial uses in 1390–1400 MHz, will have to be controlled to protect scientific uses in 1400–1427 MHz. Similar protection will have to be afforded this band by the new uses in 1427–1432 MHz.

The 1300–1427 MHz spectrum is one of many restricted bands in which the FCC's Part 15 rules permit unlicensed devices to emit only very low level emissions.

## 1427-1429 MHz

Gov: SPACE OPERATION (Earth-to-space). FIXED. MOBILE except aeronautical mobile. 722 G30

Non-Gov: SPACE OPERATION (Earth-to-space). Fixed (telemetering). Land Mobile (telemetering and telecommand). 722

FCC: Private Land Mobile (90). Satellite Communications (25).

The main mobile use in this band is air-to-ground telemetry and ground-to-air telecommand links on test ranges. Air Force data links use 1427–1435 MHz for flight testing and relay of aircraft position (see 1350–1400 MHz).

The main fixed use is by Army and National Guard tactical radio relay systems. The systems have broad tuning ranges including 200–400, 600–1000 and 1350–1850 MHz.

The 1427–1432 MHz segment will be reallocated from the government to the private sector in 1999 for uses yet to be determined. Military airborne operations in 14 geographic areas will continue in the segment until 2004.

These continuing government uses, especially the airborne stations, as well as the radio astronomy operations in 1400–1427 MHz, may make it difficult to accommodate new, commercial uses in 1427–1432 MHz.

## 1429-1435 MHz

Gov: FIXED. MOBILE. 722 G30

Non-Gov: Land Mobile (telemetering and telecommand). Fixed (telemetering). 722

FCC: Private Land Mobile (90).

The 1429–1435 MHz band is used for telecommand of missiles, experimental testing, tactical and training, light route radio relay and radar cross-section measurement.

The 1427–1432 MHz spectrum will be reallocated to the private sector in 1999 (see 1427–1429 MHz).

Navy uses include testing of shipboard electronics, control of remotely-piloted vehicles and research. Aerostat balloons, used to detect low-flying aircraft suspected of carrying drugs, operate in this band among others.

Non-government uses include antenna testing, automation and manufacturing.

## 1435–1525 MHz

Gov: MOBILE (aeronautical telemetry). 722 US78

N-G: MOBILE (aeronautical telemetry). 722 US78

FCC: AVIATION (87)

The defense and aerospace industries consider this band extremely important for aeronautical test telemetry. They use it for flight testing of manned and unmanned aircraft, missiles and space vehicles and associated communications such as range safety, chase aircraft and weather data.

NASA uses this band for uplink control of robotic aircraft in high-altitude atmospheric research. The aircraft downlink is at 2310–2360 MHz.

The aeronautical and space test uses in the 1435–1525 MHz band were believed so essential to national defense and to the commercial space industry that the United States rejected the worldwide allocation to satellite audio broadcasting in the 1452–1492 MHz.

This decision shocked nations which sought a common market and technical approach with the U.S. for 1400 MHz broadcasting. Canada has been pointedly critical of the U.S. decision and of efforts to develop digital broadcasting within the existing U.S. AM and FM radio bands.

The U.S. chose instead to allocate the band 2310–2360 MHz to satellite Digital Audio Radio Services, giving up 50 MHz that was previously available to flight testing telemetry in that band.

Nevertheless, the FCC has authorized the AfriStar 1 satellite, on an experimental basis only, as part of the WorldSpace system to broadcast in 1467–1492 MHz to developing countries.

WorldSpace will offer 216 near-CD quality channels, of which 90 will be devoted to pay audio services. The remainder will be offered to international broadcast stations, which will uplink to the satellites in 7.05–7.075 GHz. Service will begin in 1998.

WorldSpace will broadcast to special "StarMan" radios, capable of solar power operation and text readout. The radios will connect to other devices such as portable computers, for E-mail, fax and image display.

"The market for the WorldSpace service is enormous," according to the company, "made up of 3.5 billion people that are presently underserved by the poor sound quality of shortwave radio, or coverage and program limitations of AM and FM services."

The FCC apparently will not permit WorldSpace to broadcast to the United States—there is no broadcasting allocation in this band for the U.S.—but the company said that with a suitable receive antenna, listeners in the U.S., Europe or Russia could pick up the signal.

WorldSpace counsel Martine Rothblatt told us, "WorldSpace is the only worldwide licensed satellite DAB system, which also provides digital videocasting as an ancillary service. The WorldSpace system is licensed pursuant to the international WARC-92 treaty which established a worldwide primary frequency allocation, accessible immediately, in the 1467-1492 MHz band.

"The WorldSpace system is licensed by four governments—the U.S. FCC for AfriStar 1, the Government of Ghana for AfriStar A, the Government of Trinidad for CaribStar 1, and the Government of Australia for AsiaStar 1. In addition, numerous countries in all parts of the world have formally sent to the International Telecommunication Union Letters of Association with the satellite licensings just described. These Letters of Association formally establish the legal basis for transmitting a satellite beam to a particular country."

The 1435–1626.5 MHz spectrum is one of many restricted bands in which the FCC's Part 15 rules permit unlicensed devices to emit only very low level emissions.

## 1525-1530 MHz

Gov: MOBILE-SATELLITE (space-to-Earth). Mobile (aeronautical telemetry). 722 726A US78

N-G: MOBILE-SATELLITE (space-to-Earth). Mobile (aeronautical telemetry). 722 726A US78

FCC: SATELLITE COMMUNICATION (25). Aviation (87).

This band is allocated worldwide for mobile satellite downlinks. Aeronautical telemetry may continue in the band on a secondary basis.

Complex allocations in this and adjacent bands comprise the spectrum available to the Mobile Satellite Service for geostationary satellites (GSO/MSS).

These allocations include the so-called "lower L-band" at 1525–1530, 1530–1544 and 1626.5–1645.5 MHz and the "upper L-band" at 1545–1559 and 1646.5–1660.5 MHz.

The sole GSO/MSS licensee is American Mobile Satellite Corp. (AMSC), whose $500 million AMSC-1 is one of the most powerful non-government satellites yet launched.

No other U.S. companies may provide regular GSO/MSS services, ostensibly because of lack of spectrum. AMSC is authorized to use the upper L-band. The FCC proposed to authorize it to use the lower L-band contingent on international negotiations.

Other systems (of Inmarsat, Canada, Mexico and Russia) are coordinating spectrum for MSS systems in the vicinity of North America, constraining the spectrum in the L-bands available to the U.S. for MSS.

The Aeronautical Mobile Satellite (R) Service is primary in 1545–1555 and 1646.5–1656.5 MHz. The (R) indicates that the communications concern safety and regularity of flight. The 1555–1559 and 1656.5–1660.5 MHz spectrum is available to MSS with AMS(R)S having preemptive priority over the spectrum if needed.

There are no AMS(R)S satellites, but AMSC is required to accommodate AMS(R)S use should it occur. Similarly, use of the 1530–1545 and 1626.5–1646.5 MHz spectrum requires priority access to distress and safety communications. That spectrum is used by COSPAS/SARSAT, part of the Global Maritime Distress and Safety System that replaced Morse code use by ships.

AMSC began as a consortium of competing license applicants. Today AMSC shareholders include Hughes Communications, Singapore Telecom, AT&T, and Mtel.

AMSC's service, marketed as Skycell, offers voice, data and facsimile to land, maritime and aeronautical mobile and fixed users. AMSC-1 covers the U.S. including Alaska, Hawaii, Puerto Rico, the Virgin Islands and offshore areas. AMSC can provide maritime MSS communications to vessels beyond 200 miles of U.S. coastal waters.

The huge Hughes-built satellite was launched in April 1995. AMSC's Canadian partners launched MSAT, a similar satellite.

According to Leslie Taylor Associates, geostationary and non-geostationary MSS will create yearly service revenues of $9 billion by 2001.

The 1435–1626.5 MHz spectrum is one of many restricted bands in which the FCC's Part 15 rules permit unlicensed devices to emit only very low level emissions.

## 1530-1535 MHz

Gov: MARITIME MOBILE-SATELLITE (space-to-Earth). MOBILE-SATELLITE (space-to-Earth). Mobile (aeronautical telemetering). 722 726A US78 US315

Non-Gov: MARITIME MOBILE-SATELLITE (space-to-Earth). MOBILE-SATELLITE (space-to-Earth). Mobile (aeronautical telemetering). 722 726A US78 US315

FCC: SATELLITE COMMUNICATION (25). Aviation (87).

The major use of 1530–1544 MHz is for downlinks by the International Mobile Satellite Organization (Inmarsat), providing safety, news and general communications via voice, data and fax to mobile users around the world. Uplinks are at 1626.5–1646.5 MHz.

Land Earth stations—often called Coast Earth Stations in the maritime environment and Ground Earth Stations in aviation—link the Inmarsat satellites with national and international telecommunications networks.

The stations are generally owned and operated by Signatories, organizations nominated by their countries to invest in and work with Inmarsat. Congress designated COMSAT Corp. to be the U.S. Signatory in Inmarsat.

This is part of the L-band mobile satellite spectrum (see 1525–1530 MHz). The Air Force uses 1525–1535 MHz for aeronautical telemetry on a secondary basis.

The 1435–1626.5 MHz spectrum is one of many restricted bands in which the FCC's Part 15 rules permit unlicensed devices to emit only very low level emissions.

## 1535-1544 MHz

Gov: MARITIME MOBILE-SATELLITE (space-to-Earth). MOBILE-SATELLITE (space-to-Earth). 722 726A US315

Non-Gov: MARITIME MOBILE-SATELLITE (space-to-Earth). MOBILE-SATELLITE (space-to-Earth). 722 726A US315

FCC: MARITIME (80). SATELLITE COMMUNICATIONS (25).

This is part of the L-band mobile satellite spectrum (see 1525–1530 MHz).

The 1435–1626.5 MHz spectrum is one of many restricted bands in which the FCC's Part 15 rules permit unlicensed devices to emit only very low level emissions.

## 1544-1545 MHz

Gov: MOBILE-SATELLITE (space-to-Earth). 722 727A

Non-Gov: MOBILE-SATELLITE (space-to-Earth). 722 727A

FCC: MARITIME (80). SATELLITE COMMUNICATION (25).

This band is used in the COSPAS/SARSAT system to downlink distress signals (1544.5 MHz). U.S. SARSAT payloads are carried on POES and GOES satellites (see 1670–1690 and 1690–1700 MHz). This also is part of the L-band mobile satellite spectrum (see 1525–1530 MHz).

The 1435–1626.5 MHz spectrum is one of many restricted bands in which the FCC's Part 15 rules permit unlicensed devices to emit only very low level emissions.

## 1545-1549.5 MHz

Gov: AERONAUTICAL MOBILE-SATELLITE (R) (space-to-Earth). Mobile-Satellite (space-to-Earth). 722 726A US308 US309

Non-Gov: AERONAUTICAL MOBILE-SATELLITE (R) (space-to-Earth). Mobile-Satellite (space-to-Earth). 722 726A US308 US309

FCC: AVIATION (87).

This is part of the L-band mobile satellite (MSS) and Aeronautical Mobile Satellite (R) Service (AMS(R)S) spectrum (see 1525–1530 MHz).

The 1545–1555 MHz spectrum is internationally allocated to AMS(R)S downlinks, devoted to safety and regularity of flight, with uplinks at 1646.5–1656.5 MHz.

Aviation users have pre-emptive access to AMS(R)S frequencies. MSS radios must identify emergency communications and stop transmitting upon command.

The 1435–1626.5 MHz spectrum is one of many restricted bands in which the FCC's Part 15 rules permit unlicensed devices to emit only very low level emissions.

## 1549.5-1558.5 MHz

Gov: AERONAUTICAL MOBILE-SATELLITE (R) (space-to-Earth). MOBILE-SATELLITE (space-to-Earth). 722 726A US308 US309

Non-Gov: AERONAUTICAL MOBILE-SATELLITE (R) (space-to-Earth). MOBILE-SATELLITE (space-to-Earth). 722 726A US308 US309

FCC: AVIATION (87).

This is part of the L-band mobile satellite and Aeronautical Mobile Satellite (R) Service spectrum (see 1525–1530 MHz).

The 1435–1626.5 MHz spectrum is one of many restricted bands in which the FCC's Part 15 rules permit unlicensed devices to emit only very low level emissions.

## 1558.5-1559 MHz

Gov: AERONAUTICAL MOBILE-SATELLITE (R) (space-to-Earth). 722 726A US308 US309

Non-Gov: AERONAUTICAL MOBILE-SATELLITE (R) (space-to-Earth). 722 726A US308 US309

FCC: AVIATION (87).

This is part of the L-band mobile satellite and Aeronautical Mobile Satellite (R) Service spectrum (see 1525–1530 MHz).

The 1435–1626.5 MHz spectrum is one of many restricted bands in which the FCC's Part 15 rules permit unlicensed devices to emit only very low level emissions.

## 1559-1610 MHz

Gov: AERONAUTICAL RADIONAVIGATION. RADIONAVIGATION-SATELLITE (space-to-Earth) 722 US208 US260

Non-Gov: AERONAUTICAL RADIONAVIGATION. RADIONAVIGATION-SATELLITE (space-to-Earth) 722 US208 US260

FCC: AVIATION (87).

Link 2 ("L2"), 1575.42 MHz, is one of the frequencies transmitted by the Global Positioning System (see 1215–1240 and 1350–1400 MHz). Russia's Global Orbital Navigation Satellite System uses 1606–1616 MHz, but is being moved below 1606 MHz. The band also includes antenna testing and radio altimeters among its uses.

The 1435–1626.5 MHz spectrum is one of many restricted bands in which the FCC's Part 15 rules permit unlicensed devices to emit only very low level emissions.

## 1610-1610.6 MHz

Gov: AERONAUTICAL RADIONAVIGATION. RADIODETERMINATION-SATELLITE (Earth-to-space). MOBILE-SATELLITE (Earth-to-space). 722 731E 732 733 733A 733E US208 US260 US319

Non-Gov: AERONAUTICAL RADIONAVIGATION. RADIODETERMINATION-SATELLITE (Earth-to-space). MOBILE-SATELLITE (Earth-to-space). 722 731E 732 733 733A 733E US208 US260 US319

FCC: AVIATION (87). SATELLITE COMMUNICATION (25).

The 1610–1610.6 MHz band is allocated on a co-primary basis to the Mobile Satellite Service and the Radiodetermination Satellite Service (RDSS).

Radiodetermination is "the determination of the position, velocity and/or other characteristics of an object, or the obtaining of information relating to these parameters, by means of the propagation properties of radio waves."

Geostar Corp. developed RDSS. The company enraptured FCC Commissioners by promising devices for citizens to use to signal police. The devices never materialized. Police did not endorse RDSS. None of the RDSS licensees ever built their satellites. Geostar declared bankruptcy in 1991.

Geostar did, however, conduct interim services using RDSS payloads carried on the GSTAR-3 and Spacenet-3 satellites. Other companies took over these operations from Geostar. They provide services such as ship-based credit card verification and pipeline monitoring.

RDSS mobile units transmit to the satellite in 1610–1626.5 MHz and receive in 3.7–4.2 GHz in a scheme to avoid interference to radio astronomy, the Global Positioning System (GPS), the Russian Global Orbital Navigation Satellite System and terrestrial fixed users.

The interim operations will have to stop when a new service is launched. This new service was first called the Mobile Satellite Service Above 1 GHz. The FCC now uses the term Non-Geostationary Mobile Satellite Service (NGSO/MSS). It is allocated 1610–1626.5 MHz for uplinks from user stations and 2483.5–2500 MHz for downlinks to user stations. Each satellite system will use portions of this spectrum. The FCC believes these bands could accommodate as many as five competitors.

The NGSO/MSS also is informally called the "Big LEO" service, using multi-satellite constellations in low-Earth or medium-Earth orbits. Big LEOs are among the largest and most expensive satellite projects ever attempted. The Little

LEOs in the VHF spectrum are lower in cost. They offer non-voice services only (see 137–137.025 MHz).

"The Big LEO service can offer an almost limitless number of services," the FCC said, "including ubiquitous voice and data mobile services, position location services, search and rescue communications, disaster management communications, environmental monitoring, paging services, facsimile transmission services, cargo tracking, and industrial monitoring and control.

"Domestically, this service will help meet the demand for a seamless, nationwide and eventually global communications system that is available to all and that can offer a wide range of voice and data telecommunication services. In addition to enhancing the competitive market for mobile telecommunication services in areas served by terrestrial mobile services, this new mobile satellite service will offer Americans in rural areas that are not otherwise linked to the communications infrastructure immediate access to a feature-rich communications network.

"Moreover, Big LEO systems can extend these benefits throughout the world, and can provide those countries that have not been able to develop a nationwide communication service an 'instant' global and national telecommunication infrastructure.

"This network can be used to provide both basic and emergency communications to their entire populations. Big LEO systems may prove to be a critical component in the development of the global information highway."

The FCC licensed three Big LEOs. Motorola Satellite Communications is authorized for its $3.7 billion Iridium system of 66 satellites and up to 12 in-orbit spares.

Iridium will use 1621.35–1626.5 MHz. Each satellite will have intersatellite links at 23.18–23.38 GHz between adjacent satellites in the same orbital plane and adjacent orbital planes of its constellation. Iridium is authorized to use feeder links, for communications with gateway Earth stations, at 19.4–19.6 and 29.1–29.3 GHz.

TRW is authorized for Odyssey, a $1.8 billion system of 12 satellites. Unlike Iridium, Odyssey will operate in both the 1600 and 2400 MHz Big LEO bands. Odyssey feeder links are authorized at 19.8–20.1 and 29.7–30 GHz. TRW holds a broad U.S. patent for mobile satellite service to handheld telephones from medium Earth orbits.

Loral/Qualcomm is authorized for Globalstar, a $1.5 billion system of 48 satellites and eight in-orbit spares, using both Big LEO bands. Globalstar feeder links are authorized at 5.025–5.225 and 6.875–7.075 GHz.

Although rural America is an important market segment for U.S. Big LEOs, the satellites can cover most of the Earth's land areas and will serve foreign countries under arrangements with their governments and companies.

An attractive prospect is the Big LEO "world phone" that could be used anywhere on Earth. The FCC will only authorize satellite phones capable of "global roaming" if they are equipped with position determination or other capability to prevent them from transmitting from forbidden locations. Examples of such locations could be nations that ban Big LEO communications, or radio astronomy observatories that are sensitive to interference.

The International Mobile Satellite Organization (Inmarsat) established ICO Global Communications, which promises to be a formidable competitor in the Big LEO business (see 1990–2110 MHz). ICO appears similar in design to TRW's patented Odyssey. The similarity fostered disputes over patent infringement.

The Defense Department is funding development of the Eagle Eye Tag, a wristwatch-mounted device that allows a person or object to be located on demand via Big LEO satellite anywhere in the world. Applications include tracking patients, children, executives, convicts, military personnel, shipping containers and vehicles.

The 1610–1626.5 MHz spectrum also is reserved worldwide for aeronautical navigation and directly associated terrestrial or satellite-based facilities.

Radio altimeters sweep this spectrum or emit pulses that reflect off terrain to aircraft. The sweeping altimeter determines altitude by the frequency difference between transmitted and received signals. The pulsed altimeter, used especially for high altitudes, measures the delay time of the pulse from the aircraft to the terrain and back.

The Russian Global Orbital Navigation Satellite System (GLONASS) operates in 1606–1616 MHz, similar in purpose to the GPS (see 1215–1240 and 1350–1400 MHz). GLONASS operations are being moved below 1606 MHz.

International aviation interests seek complementary operation of GLONASS and GPS. Navigation and position systems that use both satellites could be more accurate than either system could be of itself. Standards are being developed for dual-system GPS/GLONASS receivers.

The combined GPS/GLONASS system is called the Global Navigation Satellite System. It is being considered as the principal long-term means of aeronautical radionavigation in the U.S. and internationally.

However, the FCC noted in February 1996 that "GLONASS has not been incorporated into or accepted as part of the global navigation satellite system for aeronautical navigation either domestically or through the International Civil Aviation Organization, and at this time there is no date certain by which that may occur."

The 1435–1626.5 MHz spectrum is one of many restricted bands in which the FCC's Part 15 rules permit unlicensed devices to emit only very low level emissions.

## 1610.6–1613.8 MHz

Gov: AERONAUTICAL RADIONAVIGATION RADIODETERMINATION-SATELLITE (Earth-to-space). MOBILE SATELLITE (Earth-to-space). RADIO ASTRONOMY. 722 731E 732 733 733A 733E 734 US208 US260 US319

Non-Gov: AERONAUTICAL RADIONAVIGATION RADIODETERMINATION-SATELLITE (Earth-to-space). MOBILE SATELLITE (Earth-to-space). RADIO ASTRONOMY. 722 731E 732 733 733A 733E 734 US208 US260 US319

FCC: AVIATION (87). SATELLITE COMMUNICATION (25).

This is part of the Big LEO spectrum (see 1610–1610.6 MHz). Radio astronomers use the band on a co-primary basis, and they use up to 1667 MHz without allocation. Big LEO satellites may have to avoid transmitting to areas with radio astronomy facilities.

The 1435–1626.5 MHz spectrum is one of many restricted bands in which the FCC's Part 15 rules permit unlicensed devices to emit only very low level emissions.

## 1613.8–1626.5 MHz

Gov: AERONAUTICAL RADIONAVIGATION RADIODETERMINATION-SATELLITE (Earth-to-space). MOBILE SATELLITE (Earth-to-space). MOBILE SATELLITE (space-to-Earth). 722 731E 731F 732 733 733E US208 US260 US319

Non-Gov: AERONAUTICAL RADIONAVIGATION RADIODETERMINATION-SATELLITE (Earth-to-space). MOBILE SATELLITE (Earth-to-space). MOBILE SATELLITE (space-to-Earth). 722 731E 731F 732 733 733E US208 US260 US319

FCC: AVIATION (87). SATELLITE COMMUNICATION (25).

This is part of the Big LEO spectrum (see 1610–1610.6 MHz). The 1435–1626.5 MHz spectrum is one of many restricted bands in which the FCC's Part 15 rules permit unlicensed devices to emit only very low level emissions.

## 1626.5-1645.5 MHz

Gov: MARITIME MOBILE-SATELLITE (Earth-to-space). MOBILE SATELLITE (Earth-to-space). 722 726A US315

Non-Gov: MARITIME MOBILE-SATELLITE (Earth-to-space). MOBILE SATELLITE (Earth-to-space). 722 726A US315

FCC: MARITIME (80). SATELLITE COMMUNICATION (25).

This is part of the L-band mobile satellite spectrum (see 1525–1530 MHz).

## 1645.5-1646.5 MHz

Gov: MOBILE-SATELLITE (Earth-to-space). 722 734B

Non-Gov: MOBILE-SATELLITE (Earth-to-space). 722 734B

FCC: MARITIME (80). SATELLITE COMMUNICATION (25).

This band is limited to distress and safety use (see 1544–1545 MHz).

It is a restricted band in which the Part 15 rules permit unlicensed devices to emit only very low level emissions.

## 1646.5-1651 MHz

Gov: AERONAUTICAL MOBILE-SATELLITE (R) (Earth-to-space) Mobile-Satellite (Earth-to-space). 722 726A US308 US309

Non-Gov: AERONAUTICAL MOBILE-SATELLITE (R) (Earth-to-space) Mobile-Satellite (Earth-to-space). 722 726A US308 US309

FCC: AVIATION (87).

This is part of the L-band mobile satellite / Aeronautical Mobile Satellite (R) spectrum (see 1525–1530 MHz).

## 1651-1660 MHz

Gov: AERONAUTICAL MOBILE-SATELLITE (R) (Earth-to-space) MOBILE-SATELLITE (Earth-to-space). 722 726A US308 US309

Non-Gov: AERONAUTICAL MOBILE-SATELLITE (R) (Earth-to-space) MOBILE-SATELLITE (Earth-to-space). 722 726A US308 US309

FCC: AVIATION (87).

This is part of the L-band mobile satellite / Aeronautical Mobile Satellite (R) spectrum (see 1525–1530 MHz).

## 1660-1660.5 MHz

Gov: AERONAUTICAL MOBILE-SATELLITE (R) (Earth-to-space) RADIO ASTRONOMY. 722 726A 736 US309

Non-Gov: AERONAUTICAL MOBILE-SATELLITE (R) (Earth-to-space) RADIO ASTRONOMY. 722 726A 736 US309

FCC: AVIATION (87).

This is part of the L-band mobile satellite / Aeronautical Mobile Satellite (R) spectrum (see 1525–1530 MHz).

Bands in 1660–1670 MHz also are allocated for radio astronomy. The 1660–1710 MHz spectrum is one of many restricted bands in which the FCC's Part 15 rules permit unlicensed devices to emit only very low level emissions.

## 1660.5-1668.4 MHz

Gov: RADIO ASTRONOMY. SPACE RESEARCH (passive). 722 US74 US246

Non-Gov: RADIO ASTRONOMY. SPACE RESEARCH (passive). 722 US74 US246

No transmissions are permitted in this receive-only scientific band. The band is used in radio astronomy observation of the spectral lines of OH: the hydroxyl radical, the first molecule to be detected at radio frequencies (1665.402 and 1667.359 MHz). Observation of these spectral lines is important for understanding the interstellar medium and star formation in galaxies.

This and higher adjacent bands are allocated for scientific uses. DSC Communications Corp. asked the FCC to allocate parts of them for wireless local loop telephone service as an alternative to copper-based networks. It would use 12 segments in 1668–2348.5 MHz. Licenses in the service would be auctioned.

The 1660–1710 MHz spectrum is one of many restricted bands in which the FCC's Part 15 rules permit unlicensed devices to emit only very low level emissions.

## 1668.4-1670 MHz

Gov: METEOROLOGICAL AIDS (radiosonde). RADIO ASTRONOMY. 722 736 US74 US99

Non-Gov: METEOROLOGICAL AIDS (radiosonde). RADIO ASTRONOMY. 722 736 US74 US99

This band is mostly used for radio astronomy in the U.S. Radiosonde operations in the band would have to be engineered to avoid conflicting with astronomy uses.

The 1660–1710 MHz spectrum is one of many restricted bands in which the FCC's Part 15 rules permit unlicensed devices to emit only very low level emissions.

## 1670-1690 MHz

Gov: METEOROLOGICAL AIDS (radiosonde). METEOROLOGICAL-SATELLITE (space-to-Earth). 722 US211

Non-Gov: METEOROLOGICAL AIDS (radiosonde). METEOROLOGICAL-SATELLITE (space-to-Earth). 722 US211

Meteorological satellites and radiosondes (weather instruments lofted by balloons) use this band in forecasting and research.

The 1670–1675 MHz segment will be reallocated from the government to the private sector for uses yet to be determined. Most federal operations are supposed to cease in that segment by January 1999. Only uses compatible with in-band and adjacent-band operations will be allowed.

The government has stated that immediate shared use of this band is feasible if interference protection is afforded to all government uses, including radiosondes, until 1999.

Internationally, 1670–1675 MHz was designated in 1992 for aeronautical telephone service. The U.S. has declared no intention of using it for that purpose. U.S. air-ground telephone services use 454–455, 459–460, 849–851 and 894–896 MHz. Mobile satellites are also expected to provide aeronautical telephone service.

The 1670–1690 MHz band contains downlinks for the Geostationary Operational Environmental Satellite System (GOES), operated by the National Environmental Satellite, Data and Information Service of the National Oceanic and Atmospheric Administration (NOAA). GOES, in operation since 1974, is the main source of the satellite weather imagery displayed on TV newscasts.

NOAA will operate two GOES satellites to monitor the Atlantic and Pacific coasts, plus a third in standby mode. NOAA's new generation of GOES satellites, "GOES-Next", has advanced capabilities. GOES 9, the second in the GOES Next series, was launched in May 1995.

GOES views the Earth at all times to produce temperature and moisture profiles of the atmosphere above the U.S. The satellite imager can detect land features, which helps forecasters track weather systems in relation to ground position. A small-scale scan feature enables the satellite to take pictures of local weather trouble spots. Hydrologic data gathered by GOES and Doppler radars (see 2700–2900 MHz) is used in water resource allocation.

GOES also monitors the Sun and its effects on the electromagnetic environment. Changes in the Sun can dramatically affect radio communications, electric power transmission, and the safety of space crews.

On 1544.5 MHz, GOES relays the 406.025 MHz distress signals received from people, aircraft or vessels in the COSPAS/SARSAT system. Ground stations analyze these signals and alert rescue centers (see 406–406.1 MHz).

Radiosondes measure air pressure, temperature and relative humidity and transmit this information back to Earth. The data collected from radiosondes are distributed globally and used in computer weather prediction models.

About half of the world's radiosondes use this band; the others are at 400–406 MHz. Radiosonde systems in this band will be replaced with new ones designed to be compatible with the new 1999 uses.

The radiosonde ground station uses a tracking antenna that follows the balloon during all phases of flight and a receiver capable of following the radiosonde's drift in frequency. If the receiver loses the radiosonde signal, it scans the entire band looking for it.

The National Weather Service launches about 80,000 radiosondes a year. Each carries a message asking the finder to return it, postage prepaid, but few are returned. Some have surfaced on the electronic surplus market.

Federal installations that will remain indefinitely in this band include the Wallops Island, Virginia, and Fairbanks, Alaska stations that control GOES and other satellites.

The 1660–1710 MHz spectrum is one of many restricted bands in which the FCC's Part 15 rules permit unlicensed devices to emit only very low level emissions.

## 1690-1700 MHz

Gov: METEOROLOGICAL-AIDS (radiosonde).
METEOROLOGICAL-SATELLITE (space-to-Earth). 671 722

Non-Gov: METEOROLOGICAL-AIDS (radiosonde).
METEOROLOGICAL-SATELLITE (space-to-Earth). 671 722

Satellite measurement of weather data and transmission of Earth imagery uses 1690–1700 and 1700–1710 MHz. This band also is allocated for radiosondes but it is largely unavailable to them because of satellite use.

These bands are used for downlinks by the Polar-orbiting Operational Environmental Satellite (POES, formerly TIROS, the Television Infrared Observation Satellite). Its command uplink is in the 148 MHz band. POES infrared and visible-spectrum instruments transmit images to meteorologists, educators and hobbyists around the world.

NOAA normally operates at least two of these satellites in polar orbit. From their 450-mile altitude, POES satellites observe temperature, cloud cover, water, ice and snow, vegetation and other environmental parameters.

They also collect data from “platform” sources such as 401 MHz radio-equipped buoys, remote weather stations and radiosondes. They convey the data to stations in Virginia and Alaska for processing and dissemination.

The Defense Meteorological Satellite Program (DMSP) performs a similar mission for military users, offering data through the Air Force Global Weather Central system.

DMSP satellites measure wind speed over the ocean; monitor precipitation to determine mobility of forces; aid in artillery and missile targeting; forecast weather over unfriendly territory; and provide data on the space environment that affects radar and ground communications. The DMSP imager can make observations at extremely low light levels and can monitor fires.

Congress has mandated that the Department of Defense and NOAA combine their polar orbiting meteorological satellite programs. Eventually the DMSP and POES functions will be combined on one spacecraft. The combined National Polar-orbiting Operational Environmental Satellite System is expected to launch the first of its three satellites in 2006.

POES satellites function in the international COSPAS/SARSAT search and rescue system. POES relays on 1544.5 MHz emergency beacon signals received on 121.5, 243 and 406.025 MHz. Ground stations analyze these signals and alert rescue centers (see 406–406.1 MHz).

The Geostationary Operational Environmental Satellite (GOES) has some downlinks in this band (see 1670–1690 MHz). On 1690.6 MHz, GOES broadcasts the Emergency Managers Weather Information Network, an experimental public source of continuous weather data including graphics and files (see 162.0125–173.2 MHz).

Not all weather satellites are government-owned. The first privately-owned and operated weather satellite is MicroLab-1. Orbital Imaging Corp. operates this satellite for NASA and the University Corporation for Atmospheric Research (UCAR). MicroLab's NASA instrument is used to study the global distribution of lightning. The UCAR experiment derives meteorological information from the effects of the atmosphere on Global Positioning System signals (see 1215–1240 MHz).

This band and 1700–1710 MHz are also used for downlinks in the Earth Exploration-Satellite Service as permitted by international footnote 671.

The 1660–1710 MHz spectrum is one of many restricted bands in which the FCC's Part 15 rules permit unlicensed devices to emit only very low level emissions.

## 1700-1710 MHz

Gov: FIXED. METEOROLOGICAL-SATELLITE (space-to-Earth). 671 722 G118

Non-Gov: METEOROLOGICAL-SATELLITE (space-to-Earth). Fixed. 671 722

See 1690–1700 MHz for this band's meteorological uses.

Fixed government and secondary, non-government industrial microwave stations use this band, particularly where 1710–1850 MHz is crowded.

From 1985 to 1988 Motorola campaigned to permit radio local area networks in this band. The FCC eventually rejected the proposal, however. It received strong opposition from the Department of Commerce, representing meteorological satellite users.

The FCC remained open to further proposals for radio LANs. Motorola later successfully petitioned to permit these products in Digital Termination System spectrum, using an unusual licensing system (see 18.8–19.7 GHz).

The 1660–1710 MHz spectrum is one of many restricted bands in which the FCC's Part 15 rules permit unlicensed devices to emit only very low level emissions.

## 1710-1850 MHz

Gov: FIXED. MOBILE. 722 US256 G42

Non-Gov: 722 US256

The 1710–1755 MHz segment of this band will be cleared of some federal users by January 2004, for eventual reallocation to the private sector for uses yet to be determined. Some emergency, military and federal power operations will continue in this segment.

Reallocation of that segment could happen as early as 1999, at least in areas near the 25 largest U.S. cities, if replacement spectrum becomes available and if the private sector reimburses federal agencies for the costs of moving their operations.

Even so, new users of this segment would have to coordinate with federal systems that will remain in the band indefinitely. Zones that need coordination, centered on defense facilities, encompass much of the U.S. East and West coasts and considerable inland areas.

As if this obstacle were not enough, hundreds of federal fixed microwave systems operate throughout 1710–1755 MHz. Many of these are operated by federal electric power agencies and may not be reallocated to private sector use.

The FCC believes that a general-use mobile service could not be deployed in many of the most populated areas of the U.S. in 1710–1755 MHz, absent some coordination method. This would reduce the revenue the government could raise by auctioning licenses in the segment.

Nevertheless, "ITA and others in the private wireless community have our eye on government spectrum to be reallocated in the 1710–1755 MHz band as a home for advanced industrial wireless communications, such as transaction and decision processing, facsimile, snapshots, slow-and full-motion video, and remote file access," according to Mark Crosby of the Industrial Telecommunications Association (ITA).

The Air Force Satellite Control Network and the Space-Ground Link Subsystem use 1761–1842 MHz (uplinks) and 2200–2290 MHz (downlinks) to control U.S. satellites, British and NATO defense and research satellites, and Space Shuttle functions, from primary stations in Guam, Hawaii, New Hampshire, Colorado and California, and from transportable stations that provide coverage for launch and orbit operations.

Other defense uses of this band include relay systems for mobile forces, missile control, air combat training systems that compute and transmit altitude, velocity and weapons status in simulations, television from aircraft-mounted cameras for remote piloting and monitoring of civil disturbances, weapon and target scoring systems, and robot telecommand for hazardous waste cleanup.

The aerostats used to detect low-flying aircraft in drug interdiction use voice and data links in this band. The Navy operates high-speed interceptor boats that transmit video in this band to nearby warships.

This is the major government band for medium-capacity point-to-point microwave links. Among the many federal fixed-service users are the Department of Energy for control and sensing in electric utility operations; the Justice Department for FBI, drug and immigration enforcement and the Interior and Agriculture Departments for national forest and park management and emergency communications in remote areas.

The Army Corps of Engineers operates a network in this band for remote controlled hydropower stations, support for the Federal Emergency Management Agency and flood control and maintenance communications at inland waterway facilities.

The 1718.8–1722.2 MHz segment is allocated internationally on a secondary basis for radio astronomy, and is used for observation of the 1720.53 MHz hydroxyl radical spectral line.

In the U.S., radio astronomy in that segment is unprotected, but other users are encouraged to minimize potential interference to radio astronomy. Accordingly, that segment is one of many restricted bands in which the FCC's Part 15 rules permit unlicensed devices to emit only very low level emissions.

## 1850-1990 MHz

Gov: US331

Non-Gov: FIXED. MOBILE. US331

FCC: PERSONAL COMMUNICATIONS SERVICES (24). PRIVATE OPERATIONAL-FIXED MICROWAVE (94). RADIO FREQUENCY DEVICES (15).

Traditionally, this band was used for the private fixed microwave stations of state and local governments, public safety agencies, petroleum, railroad, and utility companies.

Finding that these licensees have other bands and media to use, the FCC reallocated 1850–1990 MHz to the "Emerging Technologies" category for advanced services. Other Emerging Technologies bands are 2110–2150 and 2160–2200 MHz.

Emerging Technology providers must share the bands with incumbents, and pay them to move out of the bands where necessary. The presence of incumbents adds to the cost and constrains deployment of new services.

The 1850–1990 MHz band hosts the first Emerging Technology service: Personal Communications Services (PCS). The FCC originally defined PCS as "radio communications that encompass mobile and ancillary fixed communication services that provide services to individuals and businesses and can be integrated with a variety of competing networks."

The FCC said "PCS will include small, lightweight multi-function portable phones, portable facsimile and other imaging devices, new types of multi-channel cordless phones, and advanced paging devices with two-way data capabilities. ...They will make possible the vision of communications anytime, anywhere, to anyone."

The Commission later removed the requirement that fixed PCS be offered only as an "ancillary" service. As a form of

Commercial Mobile Radio Service (CMRS), PCS may now offer fixed wireless services to the public on a co-primary basis, with or without offering mobile service. This means, for example, that PCS may offer wireless local loop, basic telephone service to fixed subscribers in competition with local telephone companies.

Other forms of CMRS, such as cellular, also are permitted this option. PCS, however, is a potentially superior wireless local loop service because of its capacious spectrum allocation.

PCS developments started at the FCC in 1989 after the publicity surrounding European deployment of telepoint public cordless telephone service. Europe needed a modern platform for indoor residential and business cordless phone service. Telepoint phones were introduced to work in these private locations and through "telepoints" or low-power base stations in public places such as train platforms.

Telepoint was based on 800 MHz digital technology and inexpensive infrastructure. It fizzled in Great Britain, one of the first nations to use it.

A telepoint user could make calls but not receive them because the system was not designed to locate users. This "one-way calling" was a marketing disaster. It was misinterpreted to mean that you could talk into a phone, but you would hear nothing.

Telepoint nevertheless attracted attention from regulators as a possible low-cost competitor to cellular. Telepoint did prove popular in several Asian nations.

PCS spawned a peculiar nomenclature that began in the contentious arena of cordless telephone standards. U.S. cordless phones at 46 / 49 MHz, and older models at 1.7 MHz, made their way into European markets and were unofficially labeled CT0. Some telecom pundits incorrectly call this product CT1.

CT1 actually is a 900 MHz European standard analog UHF phone that enjoyed limited sales. A later 800 MHz version, CT1 Plus, offered more channels and interoperability among manufacturers. Britain legalized CT0-type equipment and christened it U.K. CT1, with no relation to the pan-European CT1.

To overcome the shortcomings of these early cordless products, Britain developed a second generation standard, officially released in 1987. This standard, CT2, launched the British telepoint services at 800 MHz.

A later version of CT2 permitted interoperability and was termed CT2 CAI (Common Air Interface). An enhanced CT2 CAI is CT2 Plus; a derivative of it is used in Canada.

Competing with CT2 is the 1.9 GHz Digital European Cordless Telephone (DECT), intended to be a pan-European standard. The Swedish telephone manufacturer Ericsson offered CT3, a proprietary DECT-like system.

Telepoint's decline accelerated when British regulators established the Personal Communications Network (PCN), a 2 GHz portable phone service more sophisticated than telepoint. PCN was intended to fully compete with cellular and wireline local telephone services. So far the British PCN carriers Orange and One-2-One appear successful.

The FCC adopted the term PCS to encompass PCN, telepoint (if anyone cared to offer it), and other services. Excited by telepoint and PCN, more than 200 firms applied for experimental PCS licenses. This prompted the FCC to begin lengthy proceedings to formalize PCS.

The FCC announced its band plan for PCS in 1994. It incorporates allocations for licensed PCS carriers and for unlicensed PCS devices. The total amount of spectrum allocated to licensed Broadband PCS is 120 MHz. The band plan provides for three 30-MHz licenses (blocks A, B and C) and three 10-MHz licenses (blocks D, E and F).

The A and B blocks (1850–1865 / 1930–1945 and 1870–1885 / 1950–1965 MHz, respectively) are licensed in 51 Major Trading Areas (MTAs).

These are metropolitan regions identified by Rand McNally based on geography, population and economic measures. For duplexing, each block is divided into two equal segments as shown.

The C (1895–1910 / 1975–1990 MHz), D (1865–1870 / 1945–1950 MHz), E (1885–1890 / 1965–1970 MHz) and F (1890–1895 / 1970–1975 MHz) blocks are licensed within 493 smaller Basic Trading Areas (BTAs).

Incumbent microwave licensees retain primary status in this band until the incumbent moves to another band or medium, usually after negotiation with the PCS licensee. If negotiations fail the PCS licensee may relocate the microwave licensee involuntarily.

The PCS industry has repeatedly complained that some incumbents negotiate in less than good faith, and that FCC action to protect the rights of PCS licensees is required.

Incumbents report otherwise. According to Raymond Kowalski, an attorney representing microwave interests, "The PCS industry has ignored or distorted the relocation rules, while publicly declaring themselves to be victims of the tactics of incumbents."

The stakes in these negotiations are high, not only because PCS facilities require substantial investment, but because PCS licenses themselves can be unbelievably expensive to obtain. The FCC auctions PCS licenses.

FCC Chairman Reed Hundt declared the PCS auctions as "one of the most important sales of public property since Napoleon sold almost half of the west of this country to Jefferson's United States."

Based on their experiments, three companies received early assurance of MTA license award in this band under the FCC's Pioneer's Preference program. Awardees are Omnipoint Corp. of Colorado Springs, Colo., for New York; Cox Enterprises, for Los Angeles and San Diego; and American Personal Communications (APC), serving the Washington D.C./Baltimore area.

APC, which became a part of the Sprint PCS venture, was the first PCS carrier to begin revenue operations. Marketing under the name Sprint Spectrum, it attracted 80,000 subscribers in its first five months on the air.

The FCC auction for the 99 regional Broadband PCS licenses concluded March 13, 1995. Including payments by the three Pioneer's Preference awardees of more than $701 million, the total revenue from the A and B blocks was more than $7.7 billion, declared as the largest auction in history by the Guinness Book of World Records.

Yet the C-block auction, ostensibly intended for entrepreneurs and small businesses, raised more than $10 billion when it concluded May 6, 1996. The FCC later reauctioned several C-block licenses after the high bidders failed to make necessary payments.

Winning bidders in the A and B block auctions included AT&T, Sprint and PCS PrimeCo, a partnership of regional Bell companies. The most prominent winner in the C-block auction is NextWave Personal Communications, a startup backed by PCS equipment suppliers and Asian steel and utility interests. The first auctioned PCS system to begin revenue operation started in Honolulu in February 1996, operated as VoiceStream by Western Wireless Corp.

The Personal Communications Industry Association forecasts an upbeat PCS future. It expects more than 15 million Broadband PCS subscribers by 2000, and almost 40 million by 2005. The FCC's Hundt predicts that within 10 years, half the people in the U.S. will have wireless phones.

Unlike analog cellular telephony, where carriers adhere to uniform nationwide standards, PCS will be available in several technological flavors depending on the digital standard chosen by carriers in each market.

Leading PCS standards include Time-Division Multiple Access (TDMA) approaches such as PCS-1900, based on the European Global Standard for Mobile Communications (GSM); IS-136, based on the U.S. IS-54 TDMA cellular standard; the Personal Access Communications System (PACS), based on Bellcore and Japanese telepoint-like technology; and Digital Cordless Telephone-U.S. (DCTU) a DECT derivative. PCS variants of Code-Division Multiple Access (CDMA) include J-008, based on the IS-95 U.S. CDMA standard; Broadband CDMA developed by Interdigital Communications Corp.; and a hybrid standard developed by Omnipoint.

The FCC established a separate service, Narrowband PCS, in 901–902, 930–931 and 940–941 MHz. It is used for data and paging, including digital voice messaging.

To provide for indoor, low power and short-range voice and data communications, the FCC allocated 1910–1930 MHz to unlicensed PCS devices. This allocation was prompted by Apple Computer, Inc., which envisioned unlicensed Data-PCS as part of the PCS family of services; and by the Wireless Information Networks Forum (WINForum), an association of manufacturers.

These petitioners estimated that at least 40 MHz should be allocated to unlicensed PCS. The FCC eventually divided 1910–1930 MHz into two 10 MHz portions–20 MHz total, half of the estimate–corresponding to two general categories of unlicensed devices. The 1910–1920 MHz segment is devoted to "asynchronous" applications such as wireless data communications. The 1920–1930 MHz segment is for "isochronous" uses such as cordless telephones. The FCC requires these devices to obey a spectrum resource sharing scheme.

The unlicensed 1910–1930 MHz segment contains the fewest incumbent fixed users of the entire 1850–1990 MHz band. Nevertheless, the industry must collect millions of dollars from manufacturers of unlicensed devices and use it to compensate incumbents for relocation.

Some of the money is assessed as a fee on each unlicensed PCS product. The Commission designated UTAM Inc. (Unlicensed PCS Committee for Microwave Transition and Management) to manage these funds.

Relocation is critical for unlicensed data at 1910–1920 MHz. Unlike desktop wireless networks, Data-PCS is envisioned to be portable or "nomadic" in use. Users may operate Data-PCS units at any location, where they could represent an interference threat to incumbents.

The expense, delay and difficulty of relocating fixed systems led manufacturers to seek another band for nomadic data: 2390–2400 MHz, a band allocated to the Amateur Radio Service. The FCC eventually made the band available to unlicensed PCS for data communications.

In FCC testimony, WINForum stated "While this action restores the full 20 MHz WINForum estimated was needed for those [data] technologies, the allocated spectrum unfortunately was not in a contiguous band, which may lead to interoperability challenges, limited functionality, or increased device costs. ...[O]f the original WINForum spectrum estimate, an additional 10 MHz is immediately necessary for isochronous devices."

The FCC has not promised to allocate more spectrum in this band for unlicensed devices, however.

Relocation is not an absolute, immediate necessity for products that operate under the control of a fixed base station, such as cordless telephone systems. These "non-nomadic" products may be coordinated in frequency and location so as not to interfere with incumbent stations.

Such coordination imposes funding, technical and legal requirements on non-nomadic products and suppliers that constitute de facto licensing. UTAM Inc. is the coordinator.

Products must only operate at locations known to UTAM until incumbents have moved out of 1910–1930 MHz. Manufacturers are building in safeguards to prevent a product moved away from its coordinated location from working when it is reactivated, unless UTAM approves.

At this writing, no unlicensed data PCS devices have appeared on the market. Unlicensed voice PCS is available in cordless phones for business and institutional use.

The FCC no longer grants experimental licenses for use of 1910–1930 MHz, except for applications involving unlicensed PCS. According to UTAM, "particularly during the early period, public perception of the utility and reliability of UPCS products is critical. Random shut-downs or microwave incumbent complaints (either of which could result from the testing of equipment other than UPCS devices) could cause irreparable harm and prevent this industry from bringing its full benefits to the public."

PCS in the United States eventually may form part of the Future Public Land Mobile Telecommunications System (FPLMTS), a worldwide PCS network planned by the International Telecommunication Union. The ITU has not formally allocated spectrum to FPLMTS, but has identified 1885–2025 and 2110–2200 MHz for the service.

## 1990-2110 MHz

Gov: US90 US111 US219 US222

Non-Gov: MOBILE. FIXED. US90 US111 US219 US222 NG23 NG118

FCC: AUXILIARY BROADCAST (74) CABLE TELEVISION (78)

TV stations use this band for Broadcast Auxiliary services (BAS): electronic news gathering (ENG), studio-transmitter links and intercity relay. ENG is the predominant use of 1990–2110 MHz. These mobile operations include news vans and aircraft that feed receive sites that relay the signal to the television station.

The arrival of Digital TV may exacerbate band congestion (see 470–512 MHz). TV stations will conduct "dual-mode" (ordinary NTSC TV plus DTV) auxiliary operations, increasing the burden on this spectrum.

Other TV auxiliary bands include 6.425–6.525, 6.875–7.125 and 12.70–13.25 GHz. TV broadcasters also are concerned about future interference to this band from Personal Communications Services in 1850–1990 MHz.

NASA's Tracking and Data Relay Satellite System (TDRSS) uses 2025–2110 and 2200–2300 MHz to support space missions. TDRSS relays communications between ground stations and low-orbiting spacecraft, including manned spacecraft, when the spacecraft are over the horizon and not capable of directly using ground stations.

TDRSS also operates in 13.4–14.05 GHz and 14.5–15.35 GHz. It is equipped with C-band (3.7–4.2 and 5.9–6.4 GHz) transponders leased to commercial users.

The primary band for commanding spacecraft is 2025–2120 MHz. The return link is 2200–2300 MHz. Within these bands are the NASA Deep Space Network uplink at 2110–2120 and downlink at 2290–2300 MHz. On a non-interference basis, the forthcoming Space Imaging remote sensing satellite will use 2025–2110 MHz for its tracking, telemetry and control uplink (see 8.025–8.175 GHz).

The 1992 World Administrative Radio Conference allocated 1970–1980 MHz (uplink) and 2160-2170 MHz (downlink) in Region 2 (the Americas) and the 1980–2010 MHz (uplink) and 2170–2200 MHz (downlink) bands worldwide to the mobile-satellite service (MSS).

In the U.S., however, the 1850–1990 MHz band is allocated to Personal Communications Services (PCS). "We anticipate that PCS will use spectrum intensively, thereby bringing into question the feasibility of MSS," the FCC said. "Therefore, it does not appear to be practicable to make a domestic allocation of 2 GHz spectrum for MSS that is consistent with the international allocation without jeopardizing the availability of spectrum for PCS."

With this constraint, and in an effort to be as consistent as possible with the worldwide MSS allocations, the FCC proposed to reallocate 1990–2025 MHz (uplink) and 2165–2200 MHz (downlink) to MSS.

As a part of this action, the FCC is considering proposals for new MSS services such as those from Personal Communications Satellite Corp. (PCSAT), and CELSAT. PCSAT, a subsidiary of American Mobile Satellite Corp. (AMSC), plans to launch an $885 million MSS system.

AMSC already operates a MSS system (see 1530–1535 MHz), but believes that its first-generation system will be saturated with customers, making PCSAT necessary. CELSAT would be a hybrid of satellite and terrestrial PCS, providing voice, data and video communications.

Another potential MSS user of this spectrum was established by the International Mobile Satellite Organization (Inmarsat). This personal phone satellite system, dubbed "the world's phone," expects to serve travelers who roam outside areas of cellular networks; satellite-only users; general aviation aircraft and small vessels; and fixed installations in rural and remote areas.

The $2.8 billion system would use 12 satellites in intermediate circular (not geostationary) orbits. It will be offered by a separate, London-based corporation, ICO Global Communications, in which Inmarsat is the largest of many shareholders. ICO plans to start service in 1999.

Sharing of MSS and the BAS in this band would be difficult. The FCC proposed to relocate BAS to 2110–2145 MHz. MSS would have to bear the costs of this relocation. But BAS may not be able to share with microwave incumbents in 2110–2145 MHz—so MSS would have to relocate microwave users to another band as well.

"While it might appear costly to require MSS users of the 1990–2025 MHz band to pay for relocating these incumbents in addition to relocating BAS users," the FCC said, "the costs of relocating them would be significantly mitigated by the fact that most of the microwave links at 2110–2145 MHz form one-half of a duplex system, with the paired links operating at 2160–2195 MHz.

"Because we are also proposing to allocate the 2165–2200 MHz band for MSS, MSS providers would be relocating duplex systems from 2115–2150 MHz when they relocate the corresponding paired links at 2165–2200 MHz, thereby largely clearing the 2115–2145 MHz band. Thus, MSS providers would be likely to largely clear fixed microwave incumbents at 2110–2145 MHz irrespective of whether that band is allocated for BAS."

## 2110-2150 MHz

Gov: US111 US252

Non-Gov: FIXED. MOBILE. US111 US252 NG23 NG153

FCC: DOMESTIC PUBLIC FIXED (21). PRIVATE OPERATIONAL-FIXED MICROWAVE (94). PUBLIC MOBILE (22).

The 2110–2130 MHz segment, paired with 2160–2180 MHz, has long been used in telephone Point-to-Point Microwave Service. It became popular for connections between cell sites and cellular telephone switching offices.

In proposing more allocations for Mobile Satellite Service (MSS), the FCC would require MSS providers to relocate Broadcast Auxiliary stations in 1990–2025 MHz to 2110–2145 MHz (see 1990–2110 MHz).

NASA's Deep Space Network uses the 2110–2120 MHz segment for spacecraft uplinks in the Space Research Service. Downlinks are at 2290–2300 MHz.

However important 2110–2120 MHz may be to NASA, the FCC considers that segment underutilized. "The 2110–2120 MHz band is presently allocated for fixed and mobile use. We note that there are few incumbent users on these frequencies," the Commission said in an examination of possible spectrum for the Public Safety Radio Services.

"Thus, we believe that these incumbents could either share spectrum with public safety agencies or be cleared [out] at a relatively low cost (e.g. $25,000 per link or less). This band could support 800 simplex voice channels, or 400 repeater channels, or one TV channel and 160 simplex voice channels, or other combinations," the FCC said.

The 2130–2150 MHz segment, paired with 2180–2200 MHz, is devoted to Private Operational Fixed Service licensees such as state and local governments, railroads and utilities. Many of these links will migrate to higher bands.

The 2110–2150 MHz band is one of three reallocated to "Emerging Technologies" for advanced services. The other two bands are 1850–1990 MHz, allocated to Personal Communications Services; and 2160–2200 MHz.

## 2150-2160 MHz

Gov: None.

Non-Gov: FIXED. NG23

FCC: MULTIPOINT DISTRIBUTION (21). PRIVATE OPERATIONAL-FIXED MICROWAVE (94).

The Multipoint Distribution Service (MDS) was intended for data transmission. The band became used for two channels of video when the FCC extended the high end of the band to 2162 MHz (creating room for two video channels, MDS 1 and 2) in the top 50 metropolitan areas.

This created the original microwave pay-TV service that evolved into "wireless cable". The two-channel MDS was hard-hit when cable TV arrived in many communities with a greater number of channels.

Outside major markets channel 2A (2156–2160 MHz) can be used for voice and data, being too small for a 6 MHz video channel. The 2150–2160 MHz band also is available to the Private Operational Fixed Service.

In 1983 the FCC launched the modern era of wireless competition to cable TV when it established the multichannel MDS (see 2500–2655 MHz).

## *2160-2200 MHz*

Gov: None.

Non-Gov: FIXED. MOBILE. NG23 NG153

FCC: DOMESTIC PUBLIC FIXED (21). PRIVATE OPERATIONAL-FIXED MICROWAVE (94). PUBLIC MOBILE (22).

This is one of three bands the FCC reallocated from traditional fixed and mobile uses to the new category of "Emerging Technologies" for future advanced services. The other two bands are 1850–1990 and 2110–2150 MHz.

The FCC proposed to reallocate the 2165–2200 MHz segment to downlinks in the Mobile Satellite Service (MSS), paired with proposed uplink band 1990–2025 MHz (see 1990–2110 MHz).

Incumbent licensees continue to operate microwave systems in these bands. The FCC proposed that MSS providers pay to move incumbents from this and other Emerging Technologies bands to higher bands.

Cellular carriers use the Point-to-Point Microwave Service in 2160–2180 MHz to connect base stations. As these systems add smaller cells, the long-range propagation of this 2 GHz band will not be as desirable. Future cellular use should shift to higher bands and shorter path lengths.

## 2200-2290 MHz

Gov: FIXED (Line of sight only). MOBILE (Line of sight only including aeronautical telemetering, but excluding flight testing of manned aircraft). SPACE RESEARCH (space-to-Earth) (space-to-space). SPACE OPERATION (space-to-Earth) (space-to-space). EARTH EXPLORATION-SATELLITE (space-to-Earth) (space-to-space) 750A US303 G101

Non-Gov: US303

As its detailed allocation implies, this government band supports tracking, telemetry and control for space programs, conventional fixed microwave systems and military aviation, tactical and training operations.

The government operates several satellite control networks in the intelligence, defense and civil sectors.

The National Reconnaissance Office operates classified systems including KH-11 film-return satellites, the so-called KH-12 real-time video satellite, Satellite Data System relay satellites, immense Lacrosse radar imaging satellites and other intelligence-gathering spacecraft.

The Air Force Satellite Control Network (AFSCN) is the largest network of its type in the government, operating more than 80 communication, navigation, missile warning and meteorological satellites as well as classified missions for the Department of Defense, NASA and the United Kingdom. The principal AFSCN operations center is at Falcon Air Force Base in Colorado, and is connected with many other domestic and international installations.

Key missions controlled by AFSCN include the Defense Meteorological Satellite Program (see 1690–1700 MHz);

the Global Positioning System (GPS, see 1215–1240 MHz); the Defense Support Program (DSP, see 7.25–7.3 GHz) and the Military, Strategic and Tactical Relay satellite (Milstar, see 20.2–21.2 GHz).

The Army Satellite Communications Network (ASCN) shares with AFSCN the control of the Defense Satellite Communications System (DSCS, see 7.25–7.3 GHz).

Similarly, the Naval Satellite Control Network (NSCN), with operations center at Point Mugu, California, shares control responsibility for the Fleet Satellite Communications System (FLTSATCOM), Leasat and the UHF Follow-On (UFO) satellites (see 235–267 MHz).

Some Navy and Air Force surveillance programs, which already have launched classified experimental satellites, will be consolidated into a future Space Based Wide Area Surveillance System for signals intelligence.

From the Blossom Point Tracking Facility in Maryland, the Naval Research Laboratory operates its own satellite control network for classified and scientific missions.

The largest civil satellite control network is the NASA Space Network, operated from White Sands, New Mexico. NASA uses 2200–2290 MHz intensively for space operations and research. It is a key downlink band for scientific and engineering data from spacecraft, including the Space Shuttle, via the Tracking and Data Relay Satellite System (TDRSS, see 1990–2110 MHz).

The National Oceanic and Atmospheric Administration (NOAA) Satellite Control Network supports low-Earth polar and geostationary satellites (see 1670–1690 and 1690–1700 MHz).

Another civil program is Landsat, the only non-military U.S. remote-sensing satellite system operating full-time. Since 1972, Landsat satellites have photographed most of the Earth's surface.

The Earth Observation Satellite Corp. (EOSAT) operates the current Landsats for the government. To be launched in 1998, the Landsat 7 satellite is expected to provide fast access to low-cost color and black-and-white imagery and to stimulate numerous commercial uses of its data.

NASA has responsibility for procuring and launching Landsat 7. NOAA will operate it and the Department of the Interior will store and distribute the data, for use in agriculture, mining and oil exploration, environmental and land use monitoring, mapping and water management.

This band also provides telemetry in support of airborne and seaborne target tracking and scoring, flight testing, intership data relay, missile range operations, and for radar-equipped balloons (aerostats) tethered along the southern border and Caribbean areas for use in drug interdiction.

This is one of the main bands for air-to-ground video transmission on military test ranges. The video can display missiles and other aircraft being tested. Air-to-ground video also is transmitted at 1710–1850 MHz and 4.4–4.99 GHz.

The 2200–2300 MHz spectrum is one of many restricted bands in which the FCC's Part 15 rules permit unlicensed devices to emit only very low level emissions.

## *2290–2300 MHz*

Gov: FIXED. MOBILE except aeronautical mobile. SPACE RESEARCH (space-to-Earth) (deep space only).

Non-Gov: SPACE RESEARCH (space-to-Earth) (deep space only).

This is a primary NASA Deep Space Network (DSN) band for telemetering data from probes outside Earth's orbit. DSN uplinks are at 2110–2120 MHz. These spacecraft also typically send telemetry in 8.4–8.5 GHz.

Using bands protected by the "deep space only" provision enables planetary spacecraft—at least five times farther than the Moon—to communicate with Earth without their signals being overwhelmed by the stronger signals of spacecraft closer to Earth.

The DSN receives extremely low signal levels from planetary spacecraft such as Cassini, Pioneer, Ulysses and Voyager, with signal transit time measured in hours. Communication requires very large antennas and ultra-low-noise, cryogenically-cooled amplifiers.

This band is used for Very Long Baseline Interferometry observations using multiple radio telescopes. The band also is important for measuring the polarization of space radio emissions. These measurements provide clues to radio sources and the fields through which their emissions pass.

The 2200–2300 MHz spectrum is one of many restricted bands in which the FCC's Part 15 rules permit unlicensed devices to emit only very low level emissions.

## 2300-2310 MHz

Gov: G123

Non-Gov: Amateur US253

FCC: Amateur (97).

This band was reallocated in 1995 from the government to the private sector for commercial uses yet undetermined.

Any new non-government uses would have to be compatible with the existing uses in and adjacent to this band. For example, the new uses might be limited to low-power, possibly unlicensed devices; or to stations with a low density of use, located in remote areas.

This band is considered essential to current and future Amateur Radio satellite, weak-signal, and data operations.

Government uses of this band have included military radar tests and enemy radar simulations; scoring of air-to-air missiles against drone targets; antenna experimentation and certification of defense navigation systems. Footnote G123 permits such uses to continue on a non-interference basis to non–government uses.

NASA is extremely concerned about the interference that new radio services in this band might cause to its Deep Space Network (DSN) for spacecraft tracking and telemetry in 2290–2300 MHz.

The 2300–2310 MHz band has developed obscure, non-commercial uses in part due to the need to protect DSN.

## 2310-2360 MHz

Gov: Mobile. Radiolocation. Fixed. 751B US276 US327 US328 G2 G120

Non-Gov: BROADCASTING-SATELLITE. Mobile. 751B US276 US327 US328

FCC: Digital Audio Radio Services

The principal worldwide band for audio broadcasting direct to the public from satellites (Broadcasting-Satellite Service (Sound) or BSS) is 1452–1492 MHz. That band, however, falls within aeronautical test telemetry spectrum in the U.S. (see 1435–1530 MHz).

As an alternative, the International Telecommunication Union allocated 2310–2360 MHz for domestic satellite audio broadcasting in the U.S. This action confounded other nations that prefer a uniform worldwide service. The ITU also allocated 2520–2670 MHz for BSS national and regional systems for community reception.

Against ferocious opposition from conventional broadcasters, the FCC formally allocated 2310–2360 MHz to Digital Audio Radio Services (DARS). "Satellite DARS will provide continuous radio service of compact disc quality for all listeners and will offer an increased choice of over-the-air audio programming," the FCC said.

(Footnote US328 requires DARS licensees to "take cognizance" of certain frequencies used for launch vehicles in this band.)

The DARS concept also encompasses digital broadcasting within the current AM and FM terrestrial radio allocations. This satellite allocation did not affect terrestrial DARS, however, which is still in development (see 88–108 MHz).

Four companies have applications for satellite DARS licenses pending at the FCC: Satellite CD Radio Inc., Digital Satellite Broadcasting Corp., American Mobile Radio Corp., and Primosphere L.P., a company formed by managers of several pop music megastars including Bruce Hornsby and Def Leppard.

The FCC ordinarily does not permit stations or satellites to be built without a license or construction permit. The Commission waived the rule to permit Satellite CD Radio to spend up to $10 million on building its satellite system.

"Given the amount of capital at stake, and the possibility that the Commission may not license CD Radio, this may be a risky proposition for CD Radio," the FCC said. "But if CD Radio wishes to commit funds to begin construction, we see no reason to prevent it from doing so."

Primosphere proposes an advertiser-supported service. The other three applicants propose to charge subscription fees. DARS satellites would be capable of additional, "ancillary" services such as high-speed broadcast data or location-based geographic information, voice mail, and alphanumeric messages on dedicated channels or multiplexed into the channels used for digital audio.

The FCC proposed to assign four 10 MHz licenses in 2320–2360 MHz. That plan would leave the 2310–2320 MHz segment unused due to the anticipated difficulty of keeping DARS satellites from interfering with Canadian fixed microwave systems. The FCC has asked for alternative proposals to use that segment in the U.S.

The plan would allow all four applicants to have licenses, without the need for auctions. The FCC has not ruled out the possibility of an additional filing window for satellite DARS license applications. Accepting additional license applications would probably lead to license auctions.

Another option would be to license less than 10 MHz to each of the four companies and auction the remainder. The FCC also could accept new applications and auction all licenses, possibly with financial credits afforded to the existing applicants in recognition of money they have already spent to develop the service.

The number of licenses does not automatically equal the number of satellites. To ensure sufficient quality to mobile customers, licensees may need to transmit the same signal over multiple satellites and employ so-called "gap-filler" terrestrial stations on satellite frequencies.

The FCC has not decided exactly where in the spectrum it will locate satellite DARS feeder links, which licensees will use to transmit programming up to their satellites. It appears likely, however, that at least some of these links will be expected to share part of 6.875–7.075 MHz.

It is not clear that any DARS receiver will be able to tune in any DARS service. It is more likely that satellite DARS will use multiple, incompatible technologies. The satellite DARS applicants have not converged on a single standard and the FCC has not asked them to do so.

NASA has used its Tracking and Data Relay Satellite System in transmission experiments with DARS license applicants (see 1990–2110 MHz).

The FCC licensed, on an experimental basis only, the WorldSpace DARS system to broadcast to foreign countries, although the licensee has stated its signal will be available over the air in the U.S. (see 1435–1530 MHz).

The 2310–2390 MHz spectrum is one of many restricted bands in which the FCC's Part 15 rules permit unlicensed devices to emit only very low level emissions.

## *2360-2390 MHz*

Gov: MOBILE. RADIOLOCATION. Fixed. US276 G2 G120

Non-Gov: MOBILE. US276

The National Astronomy and Ionospheric Center and Cornell University operate a planetary research radar at 2380 and 430 MHz as part of the Arecibo Radio Astronomy Observatory in Puerto Rico.

Radar echoes from objects such as comets, planets and the Moon contain information about surface properties, orbit and object size. This enables controllers, for example, to accurately guide spacecraft to specific positions on other planets. The megawatt radar also can detect asteroids that may potentially threaten civilization. It could detect a golf ball at the distance of the Moon.

Arecibo's main antenna, 1000 feet in diameter and 167 feet deep, forms the world's largest radiotelescope and radar station. The dish is affixed to a huge sinkhole in the hills of northwest Puerto Rico. It is now surrounded by a 50 foot high fence to screen out radio noise from the surrounding terrain. Built in 1960–63 to operate up to 500 MHz, a $25 million upgrade is enabling the observatory to make observations up to 10 GHz.

In 1974, Arecibo attempted to provide other worlds with details about Earth life. It transmitted data describing our Solar System, our species and DNA composition. The observatory's accomplishments include radar maps of Venus and observations in galaxies far beyond the Milky

Way. The first planets outside the Solar System were discovered around a pulsar using the Arecibo observatory.

Another powerful radar is NASA's Goldstone Solar System Radar at 2320 MHz and 8.51 GHz in the Mojave Desert. Scientists used it to measure Comet Hyakutake when it passed within 9.3 million miles of Earth in March 1996.

The band is otherwise used for telemetry in the flight testing of aircraft, spacecraft and missiles at U.S. military and NASA centers and for commercial space launches.

The 2310–2390 MHz spectrum is one of many restricted bands in which the FCC's Part 15 rules permit unlicensed devices to emit only very low level emissions.

## *2390-2400 MHz*

Gov: G122

Non-Gov: AMATEUR.

FCC: Radio Frequency Devices (15). AMATEUR (97).

The Air Force and Navy use this band for target identification, telemetry and certification of navigation systems at high speeds. It has been reallocated, however, to private sector use.

The Amateur Radio Service uses its 2390–2450 MHz spectrum for repeater links, satellite, amateur TV and data applications. Most amateur microwave links in 2390–2400 MHz are paired with frequencies at 2300–2310 MHz.

This primary amateur allocation of 2390–2400 MHz also is now available on a non-interference basis to Data-PCS: a class of unlicensed Personal Communications Service devices under Part 15 Rules for low-power, "nomadic" data transceivers that observe band-sharing behavior.

The other band available for Data-PCS is 1910–1920 MHz, but that band is filled with fixed microwave stations. Those stations must move to other bands over several years. The process requires an expensive, complex program administered by an industry trade group, UTAM Inc. It does not apply to this band.

The 2390–2400 and 2402–2417 MHz and 4.66–4.685 GHz bands are the first spectrum reallocated from government to private sector use as established by the 1993 Omnibus Budget Reconciliation Act (OBRA).

That legislation, prompted by findings that the federal government underutilizes some of its spectrum, requires the government to vacate or reduce operations in some bands and make them available to the FCC for reallocation.

"The Commission believes that allocating the band for unlicensed data-PCS, and providing for use of 2402–2417 MHz by Part 15 devices, will provide for the continued development and implementation of a new generation of advanced communications devices and services, such as wireless local area networks, digital cordless telephones, electronic article surveillance equipment, utility metering devices, fire and security alarm devices, and wireless bar code readers," the FCC said.

The Data-PCS allocation was initiated by a rulemaking petition filed by Apple Computer, Inc. in 1991 and facilitated by the OBRA legislation. "Before [this allocation] action," the company said in a news release, "the few radio bands available for wireless computing were overloaded with industrial transmitters or occupied by microwave stations; and, thus were unusable without massive relocation costs (estimated to be as much as $500 million) and ten to twelve years of band-clearing." [See 1850–1990 MHz.] "The new frequency allocation for Data-PCS—2390 to 2400 MHz—is shared only with ham radio operations, which are compatible with Data-PCS devices.

"As a result of the FCC's allocating the new band without further procedural or administrative delay, starting immediately, manufacturers can produce radio modems so educators and other users can set up their own wireless networks."

Microsoft Corp. told the FCC that Data-PCS "will operate in an entirely different manner than the technology we view as mobile or portable at present. ...Users of notebook and handheld computers might choose to transmit data between themselves. Ad-hoc work groups might be established as the need arises. ...Low-power communicating devices might operate as digital door locks, providing authentication and access to nomadic environments such as automobiles."

## *2400-2402 MHz*

Gov: 664 752 G123

Non-Gov: Amateur. 664 752

FCC: Amateur (97).

This band, used to an extent by military radar development, has been reallocated exclusively to the private sector for commercial uses yet to be determined. The Amateur Radio Service holds the only domestic allocation for this band as of our publication date, however.

Generally, consumer and commercial devices use the 2400–2483.5 MHz spectrum, especially ovens and spread spectrum communications products under Part 15 Rules.

Although some of the applications in the two bands are similar, the rules governing Part 15 devices in 2400–2483.5 MHz are different from those in 2390–2400 MHz. Many of the 2400–2483.5 MHz devices use spread spectrum.

In 2390–2400 MHz, the FCC requires monitoring and band-sharing ("spectrum etiquette") by Data-PCS as part of a new unlicensed regime (see 1850–1990 MHz). The rules preclude combining 2390–2400 and 2400–2483.5 MHz into a single, large Part 15 band.

The release of a wireless local-area network standard by the Institute of Electrical and Electronics Engineers should stimulate this market. Small 2400 MHz radios for portable computers are becoming available. One watt of power is sufficient for metropolitan-area links up to about 30 miles, depending on station design and installation.

Communication in 2400–2483.5 MHz must contend with interference from Industrial, Scientific and Medical (ISM) devices centered on 2450 MHz, especially microwave ovens. We estimate that there are more than 100 million microwave ovens in use in the U.S.

Medical uses include heating body tissues, magnetic resonance imaging, radiometry for cancer detection and even thawing of frozen organs for transplant.

Industrial RF heaters in this band treat food, chemicals, packaging, textiles and construction materials. Microwave lamps are used to manufacture fiber optic cable, automobile glass and headlamps, no-wax floor tiles and to dry printing on cans. The National Air and Space Museum in Washington, D.C. is illuminated by microwave lighting.

Radio astronomers searching for space signals have reportedly monitored microwave oven emissions for hours, not knowing their source was the observatory kitchen.

"The leakage limits on home microwave ovens prevent physical injury from their emissions, but these limits allow more than enough emissions to make the ovens powerful jammers of communications," according to the Northern California Packet Association.

Amateur Radio satellites use this band for digital transmission of telemetry, beacons and user services.

## 2402–2417 MHz

Gov: 664 752 G122

Non-Gov: AMATEUR. 664 752

FCC: AMATEUR (97). Radio Frequency Devices (15).

The FCC allocated this band to Amateur Radio on a primary basis and permitted unlicensed Part 15 devices to continue to share it on a non-interference basis.

## 2417–2450 MHz

Gov: Radiolocation. 664 752 G2 G124

Non-Gov: Amateur. 664 752

FCC: Amateur (97).

This military radar band was reallocated for shared use with the private sector for future commercial operations.

This is part of the 2400–2483.5 MHz spectrum used by unlicensed devices. Most of the radio noise from ovens and Industrial, Scientific and Medical (ISM) uses peaks at 2450 MHz, making this spectrum difficult for future commercial services covering large areas.

## 2450–2483.5 MHz

Gov: 752 US41

Non-Gov: FIXED. MOBILE. Radiolocation 752 US41

Unlicensed communications devices are allowed in 2400–2483.5 MHz, while Industrial, Scientific and Medical products may radiate up to 2500 MHz.

Law enforcement agencies use wireless surveillance cameras in this band. The band is favorable for fixed links in the Gulf of Mexico, away from urban oven interference.

## 2483.5-2500 MHz

Gov: RADIODETERMINATION SATELLITE (space-to-Earth). MOBILE SATELLITE (space-to-Earth). 752 753A 753F US41 US319

Non-Gov: RADIODETERMINATION SATELLITE (space-to-Earth). MOBILE SATELLITE (space-to-Earth). 752 753A 753F US41 US319 NG147

FCC: SATELLITE COMMUNICATION (25).

This is the Big LEO downlink band (see 1610–1610.6 MHz).

The band is mostly used by petroleum and broadcast auxiliary links. No more non-satellite licensing in the band has been allowed since 1985.

The FCC is concerned about potential interference to Big LEO downlinks from Instructional Television Fixed Service stations operating at 2500 MHz and Industrial, Scientific and Medical devices in 2400–2500 MHz.

This is one of many restricted bands in which the FCC's Part 15 rules permit unlicensed devices to emit only very low level emissions.

## 2500-2655 MHz

Gov: 720 US205 US269

Non-Gov: FIXED. BROADCASTING-SATELLITE. 720 US205 US269 NG47 NG101 NG102

FCC: AUXILIARY BROADCASTING (74). DOMESTIC PUBLIC FIXED RADIO (21).

The 2500–2655 and 2655–2690 MHz bands carry "wireless cable" entertainment in the multichannel Multipoint Distribution Service (MDS) and educational TV in the Instructional Television Fixed Service (ITFS).

Elevated transmitters broadcast wireless cable programming, typically satellite video networks, sports, movies and some local stations. Subscribers use directional, rooftop antennas and set-top boxes to view the signal.

Wireless cable has access to a maximum of 33 channels: 2150–2156 MHz (Channel 1); 2156–2162 MHz (Channel 2); the 20 ITFS channels A1–A4, B1–B4, C1–C4, D1–D4 at 2500–2596 MHz and G1–G4 at 2644–2686 MHz; the MDS channels 2596–2602 MHz, 2608–2614 MHz, 2620–2626 MHz, and 2632–2638 MHz (channels E1, E2, E3 and E4); 2602–2608 MHz, 2614–2620 MHz, 2626–2632 MHz and 2638–2644 MHz (channels F1, F2, F3 and F4); 2650–2656 MHz, 2662–2668 MHz and 2674–2680 MHz (channels H1, H2 and H3).

In 1992 the 2160–2162 MHz segment was reallocated to "Emerging Technologies" for future applications. Any subsequent MDS use of those 2 MHz will be secondary.

The term MDS refers both to single channel and to multichannel stations in the Multipoint Distribution Service. MDS stations on the E and F channel groups are sometimes called Multichannel Multipoint Distribution Service (MMDS) stations.

Not all channels are available to all operators in all locations. Complex regulations control filing of license applications and channel sharing among MDS and ITFS. The interference issue is sensitive in the many cases where MDS auction winners must protect incumbent MDS providers already operating in the area won at auction.

ITFS is defined as "a nonbroadcast service intended to be used primarily for the transmission of instructional and cultural material to specified receive sites for the formal education of students enrolled in accredited schools."

Educational institutions use ITFS to transmit instruction to remote locations. They very often lease their unused capacity to commercial MDS operators. The ITFS operator uses the channels during the school day while the MDS operator broadcasts in the evening TV hours. The MDS operator usually pays for construction of the ITFS facilities.

Mutually-exclusive applications for ITFS stations are resolved in a hearing by means of a point accumulation system. MDS licenses are now awarded by competitive bidding. The FCC's March 1996 auction of MDS licenses in 493 Basic Trading Areas raised more than $200 million.

The wireless cable industry is developing digital technology that will expand the number of channels in the available spectrum. Digital transmission also should enhance usable signal range.

This industry is subject to competition from newer services, notably direct broadcast satellites at 12 GHz and the Local Multipoint Distribution Service at 28 GHz, as well as from cable TV.

The FCC allows alternative uses other than wireless cable video transmission if the operator can satisfy the MDS technical rules or adequately support waivers of those rules.

MDS operators are already experimenting with novel services. The National Digital Network (NDN) of FCC license holder CAI Wireless Systems, is developing Personal Communications Services (PCS) and high-speed Internet services via MDS.

"Utilizing the Code Division Multiple Access (CDMA) approach with Direct Sequence Spread Spectrum (DS-SS) in a single hub system and a three sector antenna system, it is possible to accommodate up to 10,000 subscribers per 6 MHz of capacity," the company said. "Continued refinement of the technology could result in greater capacity. Since our entry point for the acquisition of spectrum is at a lower per population value than PCS spectrum, NDN could become the low cost provider of PCS services nationwide."

## 2655-2690 MHz

Gov: Earth Exploration-Satellite (passive). Radio Astronomy. Space Research (passive). US205 US269

Non-Gov: FIXED. BROADCASTING-SATELLITE. Earth Exploration-Satellite (passive). Radio Astronomy. Space Research (passive). US205 US269 NG47 NG101 NG102

FCC: AUXILIARY BROADCASTING (74). PRIVATE OPERATIONAL-FIXED MICROWAVE (94).

See 2500–2655 MHz for the uses of this band.

The 2686–2690 MHz segment is supposed to be for student responses to instructors in the Instructional Television Fixed Service, but it is difficult to use because of its proximity to other video channels.

Passive radiometers on satellites observe the Earth's surface and atmosphere. This technology works because matter emits electromagnetic radiation, corresponding to its temperature and surface characteristics. Dry and wet soil, for example, emit energy differently, as do ice and water.

The 2655–2900 MHz spectrum is one of many restricted bands in which the FCC's Part 15 rules permit unlicensed devices to emit only very low level emissions.

## 2690-2700 MHz

Gov: EARTH EXPLORATION-SATELLITE (passive). RADIO ASTRONOMY. SPACE RESEARCH (passive). US74 US246

Non-Gov: EARTH EXPLORATION-SATELLITE (passive). RADIO ASTRONOMY. SPACE RESEARCH (passive). US74 US246

No transmissions are permitted in this receive-only scientific band.

Its uses include observations of galactic and extragalactic radio sources, solar observations by the Air Force Radio Solar Telescope Network, and observations of soil moisture and ocean salinity and temperature. The U.S. Naval Observatory Interferometer at Green Bank, West Virginia uses the band to produce accurate position determinations.

This is one of many restricted bands in which the FCC's Part 15 rules permit unlicensed devices to emit only very low level emissions.

## *2700–2900 MHz*

Gov: AERONAUTICAL RADIONAVIGATION. METEOROLOGICAL AIDS. Radiolocation. 717 770 US18 G2 G15

Non-Gov: 717 770 US18

Airport Surveillance Radars, also called terminal radars, and military ground approach radars use this band in control of aircraft departure and landing.

The next-generation Doppler radar (NEXRAD) transmits in this band in locations where it does not conflict with aviation radars. It is a project of the Department of Defense, the Federal Aviation Administration and the National Weather Service (NWS). NEXRAD may expand into 2900–3100 MHz and 3.5–3.7 GHz.

NEXRAD uses the WSR-88D radar to detect the presence and intensity of rain, sleet and snow, to determine the speed and direction of wind within storms, and to identify conditions hazardous to aviation.

To be deployed at more than 100 sites throughout the country, NEXRAD is part of a large-scale NWS and Air Force radar modernization that promises better and earlier weather warnings. It also is angering homeowners worried about the radar's emissions and aesthetic impact.

Movie and TV celebrities became protesters after the NWS installed one of the first WSR-88Ds overlooking their Ojai, California homes. The stars appeared on the TV talk show circuit to confront NWS officials.

The opponents raised thousands of dollars to fight the radar. They told the *Los Angeles Times* that half of the funds will be used to inform other communities "that they are about to be raped by the National Weather Service."

The daily Solar Flux Index, an indicator of solar activity that affects ionospheric propagation of radio signals, is measured daily at 2800 MHz in Ottawa, Canada, and reported around the world to scientists and spectrum users.

This is one of many restricted bands in which the FCC's Part 15 rules permit unlicensed devices to emit only very low level emissions.

## 2900-3100 MHz

Gov: MARITIME RADIONAVIGATION. Radiolocation. 775A US44 US316 G56

Non-Gov: MARITIME RADIONAVIGATION. Radiolocation. 775A US44 US316

FCC: MARITIME (80).

Radar transponder beacons (RACONs) use this band as required by international treaty. RACONs operate together with shipborne navigation radars to electronically identify maritime obstructions and navigation points.

This is an expansion band for the NEXRAD radar (see 2700–2900 MHz). Military air traffic control, artillery and rocket tracking and long-range surveillance radars also use the band. An example is the Navy's Cobra Judy shipborne phased-array system that provides missile launch data in support of arms control agreements.

## 3.1-3.3 GHz

Gov: RADIOLOCATION. 713 778 US110 G59

Non-Gov: Radiolocation. 713 778 US110

The 3.1–3.6 GHz spectrum contains more defense radar investment than any other radiolocation band. Although radars in this band are on fixed, land mobile and ship platforms, most are airborne. The best known is AWACS, the Airborne Warning and Control System radar. AWACS tracks ships, aircraft and other emitters.

The radar portion of the Navy's premier fleet air defense system, AEGIS, operates in 3.1–3.5 GHz from guided missile cruisers and destroyers.

The 3.26–3.267 GHz segment is one of many restricted bands in which the FCC's Part 15 rules permit unlicensed devices to emit only very low level emissions.

## 3.3-3.5 GHz

Gov: RADIOLOCATION. 664 778 US108 G31

Non-Gov: Amateur. Radiolocation. 664 778 US108

FCC: Amateur (97).

The Amateur Radio Service shares this portion of the 3.1–3.6 GHz defense radar spectrum on a secondary basis.

The 3.332–3.339 and 3.3458–3.358 GHz segments are restricted bands in which the FCC's Part 15 rules permit unlicensed devices to emit only low level emissions.

## 3.5-3.6 GHz

Gov: AERONAUTICAL RADIONAVIGATION (Ground-based). RADIOLOCATION US110 G59 G110

Non-Gov: Radiolocation. US110

This part of the defense radar spectrum is an expansion band for the NEXRAD radar (see 2700–2900 MHz).

High-power military radars in this band are a potential interference threat to commercial uses in 3.65–3.7 GHz.

## 3.6–3.7 GHz

Gov: AERONAUTICAL RADIONAVIGATION (ground-based). RADIOLOCATION US110 US245 G59 G110

Non-Gov: FIXED-SATELLITE (space-to-Earth). Radiolocation. US110 US245

The 3.65–3.7 GHz segment will be reallocated to the private sector in 1999 for future commercial uses.

This segment, 50 MHz wide, is among the most important of the bands to be transferred from government use because of its contiguous width. Some government radar operations will continue in this segment indefinitely.

The Navy uses 3.6–3.7 GHz for air traffic control and precision approach radar on carriers. These radars will have to move lower in the band when the reallocation occurs.

Navy spectrum management has pointedly criticized this reallocation, and noted the potential for interference from its relocated radars to future commercial uses of the band. This may require guardbands to isolate the services, further reducing the spectrum available to the Navy.

The Air Force uses 3.625–3.65 GHz for countermeasures (jamming), radar experimentation, and data communication with hypersonic aircraft. This also is an expansion band for civil air traffic control radar that could not be accommodated in 2700–2900 MHz.

International telecommunications satellite downlinks use the "extended C-band" at 3.625–3.7 GHz. The government is examining the possible allocation of 3.625–3.65 GHz to government fixed satellite use. These applications will have to be considered when the FCC proposes uses for the band.

The 3.6–4.4 GHz spectrum is one of many restricted bands in which the FCC's Part 15 rules permit unlicensed devices to emit only very low level emissions.

# 3.7–4.2 GHz

Gov: None

Non-Gov: FIXED. FIXED-SATELLITE (space-to-Earth). NG41

FCC: DOMESTIC PUBLIC FIXED (21). SATELLITE COMMUNICATIONS (25). PRIVATE OPERATIONAL FIXED MICROWAVE (94).

The C-band is the workhorse of the satellite cable TV industry. This is the satellite downlink band; the C-band uplink is 5.925–6.425 GHz. Cable TV networks use the band to provide programming to cable systems and directly to customers with home satellite receivers.

The band also is used for tracking, telemetry and control of satellites while transferring them to final orbit position.

This band also delivers voice and data services. Many radio stations receive programming by C-band satellite. Satellite audio services also are available to home dish owners. The FCC permits mobile satellite data service to trucks and ships in this band.

The International Telecommunications Satellite Organization operates a global satellite system in the 4 / 6 and 11 / 14 GHz bands. Pan American Satellite and Columbia Communications Corp. also provide international voice, data and video satellite services.

This band hosted the first long-haul network for TV distribution and long-distance telephone. Satellite uses must share with these common-carrier fixed systems. Fiber optics are replacing common-carrier circuits in this band.

Terrestrial interference from the fixed service is common and vexing to satellite earth station installers. Owners must buy filters and shielding to help eliminate the interference. Because of this problem, new fixed links in this band have been kept out of many urban locations.

The FCC also opened this band to private microwave systems relocating from 2 GHz "Emerging Technologies" spectrum into higher bands. The satellite industry expects more interference from incoming fixed service stations.

The 3.6–4.4 GHz spectrum is one of many restricted bands in which the FCC's Part 15 rules permit unlicensed devices to emit only very low level emissions.

## 4.2–4.4 GHz

Gov: AERONAUTICAL RADIONAVIGATION. 791 US261

Non-Gov: AERONAUTICAL RADIONAVIGATION. 791 US261

FCC: AVIATION (87).

Most aircraft use radio altimeters in this internationally-reserved band. These devices determine altitude by emitting signals that reflect from terrain back to the aircraft. For high altitude measurement, some aircraft use pulsed altimeters (see 1610–1610.6 MHz).

The frequencies 4.202 and 6.427 GHz are available, but reportedly not yet used, for two-way satellite time signal communications.

Passive sensors on remote sensing satellites also may use this band to measure the temperature of the sea surface.

The 3.6–4.4 GHz spectrum is one of many restricted bands in which the FCC's Part 15 rules permit unlicensed devices to emit only very low level emissions.

## 4.4–4.5 GHz

Gov: FIXED. MOBILE.

Non-Gov: None

The federal 4 GHz band includes the segments 4.4–4.5, 4.5–4.8 and 4.8–4.99 GHz. They are allocated to fixed and mobile use on a shared primary basis.

The major users in this spectrum are the Departments of Defense, Energy and the Treasury. Links in 4.4–4.99 GHz connect major Army headquarters. It is the only part of the spectrum used by the U.S. military for tropospheric scatter communications.

Other military uses are video and radar bomb scoring, air-to-ground video, air combat maneuvering instrumentation, drone aircraft control, microwave links associated with land mobile radio systems and tactical relay.

Navy uses include data transfer between ships and helicopters, control of drone aircraft and developmental aeronautical systems.

The Department of Energy uses this band for telemetry and video systems and mobile communications in support of the little-known, anti-terrorist Nuclear Emergency Search Team (NEST). This organization must search for nuclear materials and respond to emergencies in any location. NEST radios tune throughout 4.4–4.99 GHz.

The Treasury Department operates tethered radar balloons (aerostats) in this band, deployed along the southern border and Caribbean areas to detect aircraft suspected of carrying drugs.

## 4.5–4.66 GHz

Gov: FIXED. MOBILE. US245

Non-Gov: FIXED-SATELLITE (space-to-Earth). 792A US245

See 4.4–4.5 GHz for the uses of this band. The 4.635–4.66 GHz segment will be reallocated to the private sector in 1997 for new commercial uses as yet undetermined.

For 14 additional years, new non-government uses in 4.635–4.66 GHz would have to accommodate government airborne operations in Washington, D.C., Baltimore,

Philadelphia, and portions of New York City, as well as the U.S. Virgin Islands and Puerto Rico.

As is the case with other bands reallocated from government to non-government use (for example, 2417–2450 MHz), the presence of government incumbents lessens the economic value of this band unless it is used for purposes that can function in a shared environment.

The FCC noted that this band has “essentially the same technical characteristics and potential uses” as the General Wireless Communications Service band 4.66–4.685 GHz.

The 4.5–5.15 GHz spectrum is one of many restricted bands in which the FCC’s Part 15 rules permit unlicensed devices to emit only very low level emissions.

## 4.66–4.685 GHz

Gov: G122

Non-Gov: FIXED. FIXED-SATELLITE (space-to-Earth). MOBILE. 792A US245

This band is allocated domestically on a primary basis for non-government Fixed Satellite Service downlinks, with use limited to international intercontinental systems. There is currently no non-Government use of this band.

This band is one of several reallocated to private sector use from federal government use. The FCC will license this band to fixed and mobile service applicants in a new General Wireless Communications Service (GWCS), regulated in Part 26 of the FCC Rules.

The GWCS marks a major departure from traditional FCC spectrum allocations practice. The Commission is not prescribing the purpose to which GWCS will be applied.

“Licensees could use this spectrum for dispatch service, point-to-point microwave, aeronautical audio/visual service, wireless local loop services, and terrestrial fixed and mobile auxiliary broadcast operations," the FCC said. "As a service category that is not limited to specific past and current uses, but is available for the implementation of future technologies, we expect that GWCS will encourage research and investment to invent, develop and market new technologies, and spur their deployment to serve consumers."

As long as the use made of GWCS falls under the broad categories of "fixed" or "mobile", the FCC will allow it—to entrepreneurs who win licenses at auction. On the other hand, the FCC forbids use of GWCS for broadcasting to the public, radiolocation and all forms of satellite services.

The FCC will license 4.66–4.685 GHz in five blocks, each of which is 5 MHz wide. One entity cannot obtain more than three blocks (15 MHz) in a geographic licensing area. The blocks are: A, 4.66–4.665; B, 4.665–4.67; C, 4.67–4.675; D, 4.675–4.68 and E, 4.68–4.685 GHz.

Geographic boundaries of licenses will be based on Economic Areas (EAs), zones established by the Department of Commerce. A licensee could aggregate licenses to offer regional or even nationwide service. The FCC will auction 875 licenses (five blocks in 175 EAs).

Within five years, GWCS licensees must offer service to one-third of the population in the licensed area. They must serve two-thirds within ten years of being licensed. Failure to meet these requirements will result in forfeiture of the license and the licensee will be unable to regain it.

A chief proponent of this allocation is In-Flight Phone Corp., for transmission of entertainment programming to aircraft. The composition of services in this band will depend, however, on resolution of the proposed rules and the complex auction process.

Not all sides are satisfied that the FCC has met its statutory responsibilities in allocating this band to GWCS. Jonathan Blake of the Association for Maximum Service

Television testified to the FCC: "in the 4 GHz band, the Commission failed to discharge its spectrum allocation duty altogether. Instead of determining which service suited the band and served the public, the Commission broadly allocated the spectrum for fixed and mobile services (including one-way, two-way, mobile, fixed, private, commercial, subscription, and non-subscription uses). The Commission left it to auctions to sort out the shifting and probably spotty composition of the band. ...

"Broadcasters had sought an allocation of the 4 GHz band for a non-mutually exclusive, non-subscription application: electronic news gathering (ENG) functions that are being crowded out of lower frequencies and provide critical support for live news and special events coverage both currently and in the future digital world. ...

"The amorphous and overbroad allocation the Commission adopted all but forbids the operation of ENG in the 4 GHz band by opening the frequencies to a miscellany of services with vastly different and mutually incompatible interference characteristics, operational modes, spectrum needs, and service traits."

The 4.5–5.15 GHz spectrum is one of many restricted bands in which the FCC's Part 15 rules permit unlicensed devices to emit only very low level emissions.

## 4.685-4.8 GHz

Gov: FIXED. MOBILE. US245

Non-Gov: FIXED-SATELLITE (space-to-Earth). 792A US245

See 4.4–4.5 GHz for the uses of this band.

The 4.5–5.15 GHz spectrum is one of many restricted bands in which the FCC's Part 15 rules permit unlicensed devices to emit only very low level emissions.

## 4.8-4.99 GHz

Gov: FIXED. MOBILE. 720 778 US203 US257

Non-Gov: 720 778 US203 US257

See 4.4–4.5 GHz for the uses of this band.

Formaldehyde is detected in interstellar clouds at 4829.66 GHz, an important spectral line in radio astronomy.

The 4.5–5.15 GHz spectrum is one of many restricted bands in which the FCC's Part 15 rules permit unlicensed devices to emit only very low level emissions.

## 4.99-5 GHz

Gov: RADIO ASTRONOMY. Space Research (passive). US74 US246

Non-Gov: RADIO ASTRONOMY. Space Research (passive). US74 US246

No transmissions are permitted in this receive-only scientific band, widely used in radio astronomy.

Astronomers use it to study the distributions of brightness of objects in our galaxy and others, and to make radio maps of interstellar clouds and supernova remnants. It is desirable for astronomy because of the low level of galactic background continuum radiation.

This is a key band for Very Long Baseline Interferometry, which links radio telescopes together across long distances. Scientists use this technique to study continental drift, rotation rate of the Earth, earthquakes and spacecraft navigation.

The 4.5–5.15 GHz spectrum is one of many restricted bands in which the FCC's Part 15 rules permit unlicensed devices to emit only very low level emissions.

# 5-5.25 GHz

Gov: AERONAUTICAL RADIONAVIGATION. 733 796 797 797A US211 US260 US307

Non-Gov: AERONAUTICAL RADIONAVIGATION. 733 796 797 797A US211 US260 US307

FCC: AVIATION (87). Satellite communication (25)

This aviation band will see new uses for satellites and terrestrial broadband communications.

The 1995 World Radio Conference allocated 5.092–5.25 GHz on a primary basis to uplinks in the fixed satellite service, for the use of feeder links for non-geostationary mobile satellites (see 1610–1610.6 MHz). Big LEO systems including Globalstar and ICO are on record with plans to use this segment.

The Microwave Landing System (MLS) offers standardized civil/military airport approach and landing services in 5–5.15 GHz. MLS provides aircraft with precise measurements including azimuth, elevation and range.

MLS was supposed to replace the Instrument Landing System in civil aviation (see 108–117.975 MHz). MLS signals are less affected by terrain, structures and weather.

The Federal Aviation Administration canceled large contracts for further development of MLS. The FAA stated that the Global Positioning System (see 1215–1240 MHz) has the greatest potential to provide precision landings.

MLS operations and installations continue, however. The eventual conversion of precision air navigation to GPS is subject to more tests and development. The FAA has stated that 5.15–5.25 GHz "must be protected from encroachment by non-aeronautical services to satisfy spectrum requirements for the pressing needs of a modern, global and compatible all-weather navigation system."

The FAA is developing other services for this band, including Terminal Doppler Weather Radar, differential GPS and Automatic Dependent Surveillance (ADS), which relays aircraft position information to controllers and other aircraft via data links. ADS is expected to bring safety improvements and time and fuel savings.

The Council of European Posts and Telecommunications and the European Telecommunications Standards Institute are developing High Performance Radio Local Area Networks (HIPERLAN). Some European nations allocated 5.15–5.25 GHz for HIPERLAN and others 5.15–5.3 GHz.

HIPERLAN unlicensed wireless LAN products would provide computer users with access to servers and printers, ad-hoc portable networks, high-definition video transmission, and automation control.

Studies of HIPERLAN's effects on MLS are important to its acceptance. HIPERLAN proponents say that tests show the two systems can co-exist.

In the U.S., two proponents petitioned the FCC for HIPERLAN-like services. The Wireless Information Networks Forum (WINForum) requested spectrum for SUPERNet, a Shared Unlicensed Personal Radio Network incorporating very high data rate communications.

"SUPERNet will support wireless local area networking and computer-to-computer communications at short distances with data rates of at least 20 million bits per second," the trade group said. "At that rate, SUPERNet will surpass the speed of many wired local area networks and will be able to transmit a 1000 page single spaced document in a little over three seconds."

Apple Computer petitioned for a National Information Infrastructure Band (NII Band). Apple previously petitioned for, and received, access to 2390–2400 and 1910–1920 MHz for unlicensed Data-PCS.

"In contrast to Data-PCS's 2 x 10 MHz allocation," Apple told the FCC, "NII Band users would have access to 300 MHz of unlicensed spectrum, with much broader latitude for sharing and usage. As a result, individual users' devices could be capable of faster access to networks and transfer of information—for example, the 24 Mbps data rates expected for HIPERLAN, or two and one-half times the nominal capacity of Data-PCS connections—because the total bandwidth available to use and share would be much greater.

"The copious bandwidth of the NII Band will permit microcellular frequency re-use for yet more spectrum capacity and efficiency. This capability translates not only into the ability to do conventional tasks more quickly, but also into the ability to do things not previously possible."

In addition to indoor and short-range data transfer, Apple envisions NII Band to be used for longer-range, community networking and wireless Internet: "With NII Band devices, it will be possible to communicate at distances on the order of 10 to 15 km or more. ...

"The capability to communicate on an unlicensed, wireless basis over such distances will open up new ways of communicating. Perhaps most importantly, it will make possible an entirely new application of wireless networking: the community network."

The Apple and WINForum petitions envision self-provided service on an entirely unlicensed basis, without the fees typical of cellular, PCS or other licensed providers.

The FCC responded in May 1996 with a proposal to provide 5.15–5.35 and 5.725–5.875 GHz for what it called "NII/SUPERNet devices [which] would provide short-range, high-speed wireless digital information transfer such as data, video and imaging," according to Thomas Derenge of the FCC Office of Engineering and Technology. "These devices would facilitate inexpensive wireless access to the

Information Superhighway and would foster the development of a broad range of new digital devices, which would particularly benefit entities such as schools, health care providers, libraries and small businesses."

To reduce interference to other services that have priority in this spectrum, the FCC proposes to limit NII/SUPERNet power and height. NII/SUPERNet devices that operate indoors or that use outdoor antennas no higher than 15 meters above the ground would not be deemed to cause interference to licensed services.

Owners of devices that exceed these values would have to stop or change usage if they cause interference to a licensed service, such as feeder links to mobile satellites. Big LEO satellite interests are deeply concerned over potential interference from NII/SUPERNet to satellite receivers.

NII/SUPERNet devices would be permitted 100 milliwatt Effective Isotropic Radiated Power. The FCC also is examining how to facilitate community network operations through NII/SUPERNet, possibly by authorizing higher power levels. This could, the FCC argued, transform part of NII/SUPERNet into a licensed, auctionable service.

The FCC said "if we were to change our current proposal in order to provide for higher power community network operations in the 5.725–5.875 GHz band, with the attendant longer range and greater interference potential of these operations, it might be appropriate to license this portion of the spectrum and, in the case of mutually-exclusive applications, use competitive bidding to award such licenses."

Apple, a booster of wireless community networks, responded that "longer range" and "higher power" were being confused; that adequate range is possible with low power if gain antennas are used; and that auctioning NII/SUPERNet licenses would be counterproductive.

Radar-based fuze sensors in air-to-air missiles use this band for target detection.

The 4.5–5.15 GHz spectrum is one of many restricted bands in which the FCC's Part 15 rules permit unlicensed devices to emit only very low level emissions. At presstime, the FCC had proposed to remove 5.15–5.35 GHz from the list of restricted bands.

## 5.25-5.35 GHz

Gov: RADIOLOCATION. 713 US110 G59

Non-Gov: RADIOLOCATION. 713 US110

Among the uses of the 5.25–5.925 GHz spectrum are airborne weather radars and military systems, such as Air Force test range instrumentation radars, the Army Patriot surface-to-air missile system radar and Navy surface search, navigation and fire control radars. The joint U.S., Japanese and European Earth Observation Satellite will use this band and 9.5–9.8 GHz.

Proposed NII/SUPERNet devices would use this band (see 5–5.25 GHz).

This is one of many restricted bands in which the FCC's Part 15 rules permit unlicensed devices to emit only very low level emissions. At presstime the FCC had proposed to remove 5.15–5.35 GHz from the list of restricted bands.

## 5.35-5.46 GHz

Gov: AERONAUTICAL RADIONAVIGATION. RADIOLOCATION. 799 US48 G56

Non-Gov: AERONAUTICAL RADIONAVIGATION. Radiolocation. 799 US48

FCC: AVIATION (87).

Weather radars of various types may be found throughout 5.35–5.6 GHz, operating on a non-conformance basis.

This is one of many restricted bands in which the FCC's Part 15 rules permit unlicensed devices to emit only very low level emissions.

## 5.46-5.47 GHz

Gov: RADIONAVIGATION. Radiolocation. 799 US49 US65 G56

Non-Gov: RADIONAVIGATION. Radiolocation. 799 US49 US65

The 5.46–5.65 GHz spectrum is available for shipborne navigation radars. See 5.25–5.35 GHz for other uses of this band.

## 5.47-5.6 GHz

Gov: MARITIME RADIONAVIGATION. Radiolocation. US50 US65 G56

Non-Gov: MARITIME RADIONAVIGATION. Radiolocation. US50 US65

FCC: MARITIME (80).

See 5.25–5.35, 5.35–5.46 and 5.46–5.47 GHz for uses of this band.

## 5.6-5.65 GHz

Gov: MARITIME RADIONAVIGATION. METEOROLOGICAL AIDS. Radiolocation. 802 US51 US65 G56

Non-Gov: MARITIME RADIONAVIGATION. METEOROLOGICAL AIDS. Radiolocation. 802 US51 US65

FCC: MARITIME (80).

Terminal Doppler Weather Radar (TDWR) uses this band for air traffic control, and to detect rapid downdrafts dangerous to aircraft at low altitudes (wind shear).

"The doppler radar virtually eliminates the threat of undetected, deadly wind shears," according to Secretary of Transportation Federico Peña.

"It enables controllers to pinpoint wind shears in arrival and departure corridors near airports, and then to guide aircraft around the trouble spots. Wind shear is the number one killer of airline passengers. ...It's preventable, and we're doing something about it by installing doppler radar at 45 of our biggest airports nationwide."

Like the NEXRAD weather radar (2700–2900 MHz), the TDWR project is a political football. Residents, concerned about radiation or aesthetics, have blocked or delayed radar construction in several cities.

This also is part of the 5.46–5.65 GHz spectrum available for shipborne navigation radars.

## *5.65-5.85 GHz*

Gov: RADIOLOCATION. 664 806 808 G2

Non-Gov: Amateur. 664 806 808

FCC: Amateur (97). Industrial, Scientific, and Medical Equipment (18).

The major federal users of this band are the Department of Defense and NASA for surface, missile and rocket tracking radars, aircraft guidance and surveillance radar, telemetry, and ground facilities to develop these systems.

Radar use includes interrogating transponders onboard craft. These devices reply on a different frequency, or they may amplify the signals reflected back to the radar.

Many radars in this band are on ships, operating across the 5.4–5.9 GHz range to track targets and for Identification Friend or Foe (IFF) interrogation.

Uses in this band include control of remotely-piloted aircraft; space and missile tracking; radar for flight and trajectory measurement at testing ranges; and other fixed and mobile radars.

FCC Part 15 rules for unlicensed devices permit field-disturbance sensors, such as door openers, to operate in the 5.785–5.815 GHz spectrum. Spread spectrum unlicensed devices may use 5.725–5.85 GHz and non-spread spectrum devices may use 5.725–5.875 GHz.

At this writing, the FCC was considering a proposal to eliminate a limit on directional antenna gain for fixed, point-to-point spread spectrum systems in this band.

This provision would apply only to non-consumer, commercial and industrial fixed systems. Point-to-multipoint, omnidirectional and multiple co-located systems transmitting the same information would not be permitted to take advantage of the relaxed antenna rules.

There could be no legal marketing of these enhanced-antenna wireless systems to the general public. The FCC believes that these measures may be necessary in order to reduce interference potential from these high-gain systems, and to reduce the chance that the public would be exposed to harmful radiofrequency energy.

European nations are adopting the 5.8 GHz spectrum as the location for automatic vehicle identification tags. This reportedly enhances prospects for an international standard. In the U.S. most of these tags are at 902–928 MHz.

Some wireless local-area networks for computers use the band for spread spectrum communications (see 902–928 and 2390–2450 MHz). The FCC also proposed to designate this band for NII/SUPERNet wireless broadband devices (see 5–5.25 GHz).

The Amateur Radio Service shares the 5.65–5.925 GHz spectrum on a secondary, non-interference basis, with 5.83–5.85 GHz allocated for amateur satellite downlinks.

# 5.85-5.925 GHz

Gov: RADIOLOCATION. 806 US245 G2

Non-Gov: FIXED-SATELLITE (Earth-to-space). Amateur. 806 US245

FCC: Amateur (97).

In addition to military radar, drone control, air combat training and Amateur Radio, international satellite systems can use this band for Fixed Satellite Service uplinks.

This band also is part of the 5.725–5.875 GHz unlicensed device spectrum, proposed as part of the NII/SUPERNet band (see 5–5.25 GHz).

Intelligent Transportation Systems (ITS, formerly Intelligent Vehicle Highway Systems) seeks to use advanced computing and communications to provide real-time information to drivers, transit riders and trip planners.

ITS has been described as a "mobile information utility", and it is expected to use a wide variety of existing commercial services such as cellular, PCS, Location and Monitoring Services (LMS) and broadcast stations.

Nevertheless, some services may require wireless systems dedicated to ITS. According to the Transportation Department, "a consensus has been reached among the government and industry partners in the ITS program that the 5850–5925 MHz band is a good choice for short-range ITS communications systems.

"The huge market for systems that will accompany ITS deployment will drive down the cost of the semiconductor technology that operates at these frequencies, which will benefit all users of the band."

The FCC was not persuaded by this claim, however, to dedicate any spectrum to ITS. "At this time the spectrum requirements for ITS and their possible impact on other services are not clear," the FCC said in May 1996. The

Commission said it would address the issue of additional ITS spectrum in future rule makings as appropriate.

The FCC has established an ITS-like category, the Transportation Infrastructure Radio Services (TIRS). So far the only TIRS service is LMS (see 902–928 MHz).

## 5.925-6.425 GHz

Gov: None.

Non-Gov: FIXED. FIXED-SATELLITE (Earth-to-space). NG41

FCC: DOMESTIC PUBLIC FIXED (21). SATELLITE COMMUNICATIONS (25). PRIVATE OPERATIONAL FIXED MICROWAVE (94).

This is the C-band satellite uplink (see 3.7–4.2 GHz).

The common carrier Point-to-Point Microwave Service used by local exchange and long-distance telephone companies shares this band. Fiber optics are replacing many of these telephone links.

Cellular carriers use the band for backbone networks. The band also is available to TV stations for studio-transmitter link use. This is one of the bands to which microwave incumbents will migrate from the 2 GHz spectrum reallocated for Personal Communications Services and other "Emerging Technologies".

## 6.425-6.525 GHz

Gov: 791 809

Non-Gov: FIXED-SATELLITE (Earth-to-space). MOBILE. 791 809 NG122

FCC: AUXILIARY BROADCAST (74) CABLE TELEVISION (78) DOMESTIC PUBLIC FIXED (21) PRIVATE OPERATIONAL-FIXED MICROWAVE (94)

The 6.425–6.525 and 6.525–6.725 GHz segments may see use for Big LEO feeders (see 1610–1610.6 MHz).

The main terrestrial uses at 6.425–6.525 GHz are TV remote pickup, a Broadcast Auxiliary use; and for the Local Television Transmission Service where common carriers provide service to TV stations, cable systems and networks.

The 6.425–7.250 GHz spectrum is available to passive sensors for ocean temperature measurements.

The frequencies 6.427 and 4.202 GHz are available but unused for two-way satellite time communications.

## 6.525-6.875 GHz

Gov: 809

Non-Gov: FIXED. FIXED-SATELLITE (Earth-to-space). 809

FCC: DOMESTIC PUBLIC FIXED (21). SATELLITE COMMUNICATIONS (25). PRIVATE OPERATIONAL FIXED MICROWAVE (94).

This is a general-purpose fixed microwave band used by governmental, industrial, communications carrier and transportation licensees.

It is the main band to which microwave incumbents will migrate from the 2 GHz spectrum reallocated for Personal Communications Services and "Emerging Technologies".

Radio astronomers have discovered a strong spectral line of the methanol molecule at 6668.518 MHz. The methanol line is an important tracer of star formation activity.

## 6.875-7.075 GHz

Gov: 809

Non-Gov: FIXED. FIXED-SATELLITE (Earth-to-space). MOBILE 809 NG118

FCC: AUXILIARY BROADCAST (74) CABLE TELEVISION (78) DOMESTIC PUBLIC FIXED (21)

The 6.875–7.125 GHz spectrum is one of the bands allocated on a primary basis to Broadcast Auxiliary services. These include electronic news gathering, remote event coverage, intercity relay and studio-transmitter links. Other primary TV Broadcast Auxiliary services bands are 1990–2110 MHz and 12.7–13.25 GHz.

The advent of Digital TV may exacerbate congestion in TV auxiliary bands (see 470–512 MHz). Many stations will have "dual-mode" (NTSC plus DTV) operations, increasing the burden on this spectrum.

Feeder links for Big LEO satellites may use this band (see 1610–1610.6 MHz). The FCC also may permit satellite Digital Audio Radio Services feeder links to share this band (see 2310–2360 MHz).

The success of such sharing arrangements is likely to depend on locating satellite feeder stations away from the fixed auxiliary receive sites of TV stations.

## 7.075-7.125 GHz

Gov: 809

Non-Gov: FIXED. MOBILE. 809 NG118

FCC: AUXILIARY BROADCAST (74) CABLE TELEVISION (78) DOMESTIC PUBLIC FIXED (21)

See 6.875–7.075 GHz for information about this band.

## 7.125-7.19 GHz

Gov: FIXED. 809 US252 G116

Non-Gov: 809 US252

The Federal Aviation Administration uses the 7.125–8.5 GHz spectrum to link air traffic and radar sites.

The Department of Energy and power agencies such as the Tennessee Valley Authority use this spectrum for

System Control And Data Acquisition for power distribution. NASA uses it for deep space communications. Army and Air Force uses include security video, electronic warfare, bomb scoring and ground forces communications.

Military and government satellite uses include the Defense Satellite Communications System (DSCS), with service links in this part of the spectrum and tracking, telemetry and control links in the 1761–1842 and 2200–2290 MHz Space-Ground Link Subsystem bands.

The geostationary DSCS satellites provide the Departments of Defense, State and other agencies with jam-resistant communications. These bands also host the Defense Support Program (DSP) satellites, which use infrared telescopes to detect missiles. DSP was successfully used in the Gulf War to detect SCUD launches from Iraq.

The Air Force will replace DSP satellites with the $22 billion Space Based Infrared System (SBIRS) of geostationary and elliptical orbit satellites. A low-altitude component, Brilliant Eyes, will track missiles in midcourse.

## 7.19–7.235 GHz

Gov: FIXED. SPACE RESEARCH (Earth-to-space). 809

Non-Gov: 809

See 7.125–7.19 GHz for information about this band.

## 7.235–7.25 GHz

Gov: FIXED. 809

Non-Gov: 809

See 7.125–7.19 GHz for information about this band.

## 7.25-7.3 GHz

Gov: FIXED-SATELLITE (space-to-Earth). MOBILE-SATELLITE (space-to-Earth). Fixed. G117

Non-Gov: None

See 7.125–7.19 GHz for information about this band.

The 7.25–7.75 GHz spectrum is one of many restricted bands in which the FCC's Part 15 rules permit unlicensed devices to emit only very low level emissions.

## 7.3-7.45 GHz

Gov: FIXED. FIXED-SATELLITE (space-to-Earth). Mobile-Satellite (space-to-Earth). G117

Non-Gov: None

See 7.125–7.19 GHz for information about this band.

The 7.25–7.75 GHz spectrum is one of many restricted bands in which the FCC's Part 15 rules permit unlicensed devices to emit only very low level emissions.

## 7.45-7.55 GHz

Gov: FIXED. FIXED-SATELLITE (space-to-Earth). METEOROLOGICAL-SATELLITE (space-to-Earth). Mobile-Satellite (space-to-Earth). G104 G117

Non-Gov: None

See 7.125–7.19 GHz for information about this band.

The Meteorological-Satellite Service allocation is used to downlink weather data.

The 7.25–7.75 GHz spectrum is one of many restricted bands in which the FCC's Part 15 rules permit unlicensed devices to emit only very low level emissions.

## 7.55-7.75 GHz

Gov: FIXED. FIXED-SATELLITE (space-to-Earth). Mobile-Satellite (space-to-Earth). G117

Non-Gov: None

See 7.125–7.19 GHz for information about this band.

The 7.25–7.75 GHz spectrum is one of many restricted bands in which the FCC's Part 15 rules permit unlicensed devices to emit only very low level emissions.

## 7.75-7.9 GHz

Gov: FIXED.

Non-Gov: None

See 7.125–7.19 GHz for information about this band.

## 7.9-8.025 GHz

Gov: FIXED-SATELLITE (Earth-to-space). MOBILE-SATELLITE (Earth-to-space). Fixed. G117

Non-Gov: None

See 7.125–7.19 GHz for information about this band.

## 8.025-8.175 GHz

Gov: EARTH EXPLORATION-SATELLITE (space-to-Earth). FIXED. FIXED-SATELLITE (Earth-to-space). Mobile-Satellite (Earth-to-space) (No Airborne Transmission) US258 G117

Non-Gov: US258

Earth Exploration-Satellite Service (EES) satellites use 8.025–8.4 GHz to deliver images and for tracking, telemetry and control downlinks.

Among U.S. users are Landsat and EarthWatch, OrbView and Space Imaging commercial remote sensing satellites. These will offer one-meter resolution images, formerly available only from intelligence satellites (see 2200–2290 MHz). EarthWatch became the first U.S. EES licensee.

"Land remote-sensing satellites can provide information that can assist in studying and understanding human impacts on the environment, managing resources, and carrying out national security functions," the FCC said. "The collected data can be used for mapping, conservation, law enforcement, environmental monitoring, and forecasting change in vegetation and other features of the Earth's surface."

The 8.025–8.5 GHz spectrum is one of many restricted bands in which the FCC's Part 15 rules permit unlicensed devices to emit only very low level emissions.

## 8.175–8.215 GHz

Gov: EARTH EXPLORATION-SATELLITE (space-to-Earth). FIXED. FIXED-SATELLITE (Earth-to-space). METEOROLOGICAL-SATELLITE (Earth-to-space). Mobile-Satellite (Earth-to-space) (no airborne transmissions). US258 G104 G117

Non-Gov: US258

See 8.025–8.175 GHz for the uses of this band.

The 8.025–8.5 GHz spectrum is one of many restricted bands in which the FCC's Part 15 rules permit unlicensed devices to emit only very low level emissions.

## 8.215–8.4 GHz

Gov: EARTH EXPLORATION-SATELLITE (space-to-Earth). FIXED. FIXED-SATELLITE (Earth-to-space). Mobile-Satellite (Earth-to-space) (no airborne transmissions). US258 G117

Non-Gov: US258

See 8.025–8.175 GHz for the uses of this band.

The 8.025–8.5 GHz spectrum is one of many restricted bands in which the FCC's Part 15 rules permit unlicensed devices to emit only very low level emissions.

## 8.4-8.45 GHz

Gov: FIXED. SPACE RESEARCH (space-to-Earth) (deep space only).

Non-Gov: None

The 8.4–8.5 GHz spectrum contains space research downlinks, with 8.4–8.45 GHz limited to spacecraft beyond Earth's orbit. This "deep space only" provision is applied to some bands because of the different radio characteristics at these distances. Uplinks to deep space receivers must use more power than uplinks to satellites closer to Earth, and downlinks from deep space have lower received power than those from closer satellites.

The 8.025–8.5 GHz spectrum is one of many restricted bands in which the FCC's Part 15 rules permit unlicensed devices to emit only very low level emissions.

## 8.45-8.5 GHz

Gov: FIXED. SPACE RESEARCH (space-to-Earth).

Non-Gov: SPACE RESEARCH (space-to-Earth).

See 8.4–8.45 GHz for the uses of this band.

The 8.025–8.5 GHz spectrum is one of many restricted bands in which the FCC's Part 15 rules permit unlicensed devices to emit only very low level emissions.

## 8.5-9 GHz

Gov: RADIOLOCATION. 713 US53 US110 G59

Non-Gov: Radiolocation. 713 US53 US110

The 8.5–10.55 GHz spectrum is used for military and civil radar, including airborne navigation; Marine Corps transportable weapon-locating radars; Navy fire control radars; Army ballistic missile defense imaging radars; and space-based radars used in remote sensing applications.

In the Mojave Desert, NASA's Goldstone Solar System Radar is used to study planetary bodies. Its transmitter can output up to 1 MW at 8.51 GHz. Pulses from the radar can take as long as 4.5 hours to return from the target.

## 9-9.2 GHz

Gov: AERONAUTICAL RADIONAVIGATION. Radiolocation. 717 US48 US54 G2 G19

Non-Gov: AERONAUTICAL RADIONAVIGATION. Radiolocation. 717 US48 US54

FCC: AVIATION (87).

Military aviation uses this band for precision approach radar. This is a restricted band in which Part 15 rules permit unlicensed devices to emit only very low level emissions.

## 9.2-9.3 GHz

Gov: MARITIME RADIONAVIGATION. Radiolocation. 823 824A US110 G59

Non-Gov: MARITIME RADIONAVIGATION. Radiolocation. 823 824A US110

Maritime navigation radars operate in 9.2–9.5 GHz. They also work with search and rescue transponders that respond when interrogated by the radar. Evacuees can grab the wand-like transponder and take it onboard survival craft.

The Air Force makes extensive use of this band for the air and ground control of unmanned aerial vehicles used in electronic warfare and reconnaissance.

## 9.3–9.5 GHz

Gov: RADIONAVIGATION. Meteorological Aids. Radiolocation. 775A 824A 825A US51 US66 US67 US71 G56

Non-Gov: RADIONAVIGATION. Meteorological Aids. Radiolocation. 775A 824A 825A US51 US66 US67 US71

This is a popular band for weather radars operated by state and local agencies, broadcasters, researchers and forecasters. Many aircraft carry weather radars in this band.

It also is the main band for shipborne navigation radars and radar transponder beacons, which identify maritime obstructions and navigation points. Coast Guard radars use the band to observe harbor and coastal traffic.

This is a restricted band in which the Part 15 rules permit unlicensed devices to emit only very low level emissions.

## 9.5–10 GHz

Gov: RADIOLOCATION. 713 828 US110

Non-Gov: Radiolocation. 713 828 US110

The Space Station radar is expected to use 9.5–9.8 GHz. Other planned uses include the joint U.S., Japanese and European Earth Observation Satellite and the Space Shuttle, to carry synthetic aperture radar in this band.

The Air Force uses this band to control unmanned aerial vehicles. It also uses this and nearby bands for wideband data links for high-performance aircraft, capable of transmitting jam-resistant data over long distances.

## 10–10.45 GHz

Gov: RADIOLOCATION. 828 US58 US108 G32

Non-Gov: Amateur. Radiolocation. 828 US58 US108 NG42

FCC: Amateur (97). Private Land Mobile (90).

The 10–10.5 GHz spectrum is available to military radar, satellite weather radar and low-power survey operations.

It is becoming more popular as an Amateur Radio band. Communication range in this spectrum has exceeded 1000 miles due to tropospheric ducting.

## *10.45-10.5 GHz*

Gov: RADIOLOCATION. US58 US108 G32

Non-Gov: Amateur. Amateur-Satellite. Radiolocation. US58 US108 NG42 NG134

FCC: Amateur (97). Private Land Mobile (90).

See 10–10.45 GHz for the uses of this band.

## *10.5-10.55 GHz*

Gov: RADIOLOCATION. US59

Non-Gov: RADIOLOCATION. US59

FCC: PRIVATE LAND MOBILE (90).

This is one of two widely-used bands for police radars. "X-band" units transmit on 10.525 GHz. The other standard band is 24.05–24.25 GHz. Field disturbance sensors for door openers and security systems also use these bands.

## *10.55-10.6 GHz*

Gov: None

Non-Gov: FIXED.

FCC: DOMESTIC PUBLIC FIXED (21). PRIVATE OPERATIONAL FIXED MICROWAVE (94).

The Digital Electronic Message Service (DEMS), a common carrier data service, and private, but similar, Digital Termination System (DTS) stations were allocated 10.55–10.68 GHz.

Few DEMS or DTS stations were built. Existing licensees may stay and can expand their systems, but the FCC accepts no new DTS or DEMS license applications. The FCC converted this spectrum for conventional fixed use.

Most of the use of this spectrum is for fixed microwave links, connecting cell sites with the central office of a cellular system, for example. Microwave systems will migrate from the 2 GHz "Emerging Technologies" spectrum to other bands including this one.

## 10.6-10.68 GHz

Gov: EARTH EXPLORATION-SATELLITE (passive). SPACE RESEARCH (passive). US265 US277

Non-Gov: EARTH EXPLORATION-SATELLITE (passive). FIXED. SPACE RESEARCH (passive). US265 US277

FCC: DOMESTIC PUBLIC FIXED (21). PRIVATE OPERATIONAL FIXED MICROWAVE (94).

This is part of the 10.55–10.68 spectrum intended for multipoint data services that never extensively developed. The 10.6–10.7 GHz spectrum is used in satellite-based land and ocean surface measurements. Passive sensing and radio astronomy use 10.68–10.7 GHz on a primary basis.

The 10.6–12.7 GHz spectrum is one of many restricted bands in which the FCC's Part 15 rules permit unlicensed devices to emit only very low level emissions.

## 10.68-10.7 GHz

Gov: EARTH EXPLORATION-SATELLITE (passive). RADIO ASTRONOMY. SPACE RESEARCH (passive). US74 US246

Non-Gov: EARTH EXPLORATION-SATELLITE (passive). RADIO ASTRONOMY. SPACE RESEARCH (passive). US74 US246

No transmissions are permitted in this scientific band. Its uses include remote sensing of rainfall and astronomy studies of quasars, which are among the most energetic and distant objects that astronomers can detect.

The 10.6–12.7 GHz spectrum is one of many restricted bands in which the FCC's Part 15 rules permit unlicensed devices to emit only very low level emissions.

## 10.7-11.7 GHz

Gov: US211

Non-Gov: FIXED. FIXED-SATELLITE (space-to-Earth). 792A US211 NG41 NG104

FCC: DOMESTIC PUBLIC FIXED (21). PRIVATE OPERATIONAL FIXED MICROWAVE (94).

Telephone companies used this fixed band for analog and digital telephone and video transmission. The links are being replaced with fiber optics. This is also one of the bands to which incumbents will migrate from 2 GHz spectrum reallocated to "Emerging Technologies."

The band is available for TV studio-transmitter links and for satellite feeder links. The American Mobile Satellite Corp. uses 10.75–10.95 and 13–13.15 GHz to feed its satellite (see 1530–1535 MHz).

The 10.6–12.7 GHz spectrum is one of many restricted bands in which the FCC's Part 15 rules permit unlicensed devices to emit only very low level emissions.

## 11.7-12.2 GHz

Gov: 837 839

Non-Gov: FIXED-SATELLITE (space-to-Earth). Mobile except aeronautical mobile. 837 839 NG143 NG145

FCC: DOMESTIC PUBLIC FIXED (21). SATELLITE COMMUNICATION (25).

This is the Ku-band satellite downlink allocation. The uplink is 14–14.5 GHz. The band is used for private networks for voice, data, facsimile and video. The Public Broadcasting Service feeds its TV stations in this band. The American Mobile Satellite Corp. monitors and controls its satellite in 12 / 14 GHz (see 1530–1535 MHz).

The FCC proposed to limit use of this band by the Fixed Satellite Service in ITU Region 2 to national and subregional systems and to provide that fixed-satellite transponders operating in the band may also be used for broadcasting-satellite transmissions.

Such uses include subscription services offered by PrimeStar and Alphastar via spacecraft that otherwise are devoted to non-broadcast uses. The FCC describes these as "medium-power direct-to-home" services.

This is another band available for Broadcast Auxiliary Service studio-transmitter links and TV remote pickup.

The 10.6–12.7 GHz spectrum is one of many restricted bands in which the FCC's Part 15 rules permit unlicensed devices to emit only very low level emissions.

## 12.2-12.7 GHz

Gov: 839 843 844

Non-Gov: FIXED. BROADCASTING-SATELLITE. 839 843 844 NG139

FCC: INTERNATIONAL PUBLIC (23). PRIVATE OPERATIONAL-FIXED MICROWAVE (94). DIRECT BROADCAST SATELLITE SERVICE (100).

The FCC reallocated this band to downlinks for the direct broadcast satellite (DBS) service from terrestrial fixed microwave services in 1982. The action prompted large losses in the fixed microwave supply business.

DBS ventures of the 1980s lost huge sums of money without providing service. The U.S. DBS industry resurrected itself with the launch of dedicated DBS satellites by Hughes. They serve DirecTV, a unit of GM Hughes Electronics; and U.S. Satellite Broadcasting (USSB), a subsidiary of Hubbard Broadcasting.

The DirecTV uplink in Castle Rock, Colorado and USSB's Oakdale, Minnesota uplink feed the Hughes satellites at 17.3–17.8 GHz. The wide orbital spacing, high power and SHF operation of DBS enable consumers to use small, unobtrusive antennas to receive the signals.

At this writing Hughes operates three DBS satellites. The second, DBS-2, was launched in August 1994. It carries an unusual payload: SpaceArc, Archives of Humanity. SpaceArc contains a recording of 40,000 messages from people in 45 countries, including vice-presidents Dan Quayle and Al Gore; a documentary on events of 1993; and music of artists including Jimi Hendrix and Peter Gabriel.

The third domestic DBS competitor to begin operations was the satellite equipment manufacturer and program distributor EchoStar. Its EchoStar-101 satellite offers the DISH Network, unveiled in March 1996.

EchoStar will increase capacity with the launch of a second DBS satellite. The company won the FCC license for the second satellite at auction for $52 million. The license authorizes an orbital location that provides service to most of the U.S. except for a part of the northeast.

MCI won the final available nationwide DBS license, for which it bid $682.5 million. MCI plans to launch its first and second DBS satellites in 1997 and 1998.

MCI indicated that it will provide "more than television," including services such as software transmission. The MCI DBS service reportedly will cost $1 billion to build, in addition to the auction payment.

DBS programming includes the same national networks available on cable TV, plus pay-per-view movies, CD-quality audio channels and data services. The FCC has granted frequency and orbital assignments to several other DBS companies. Internationally, DBS is designated as the Broadcast Satellite Service (BSS).

The competitive potential of DBS earned it the nickname "Death Star" by the cable TV industry.

The 10.6–12.7 GHz spectrum is one of many restricted bands in which the FCC's Part 15 rules permit unlicensed devices to emit only very low level emissions.

## 12.7-12.75 GHz

Gov: None

Non-Gov: FIXED. FIXED-SATELLITE (Earth-to-space). MOBILE. NG53 NG118

FCC: AUXILIARY BROADCASTING (74). CABLE TELEVISION RELAY (78). PRIVATE OPERATIONAL-FIXED MICROWAVE (94).

The main terrestrial users of the 12.7–12.75 and 12.75–13.25 GHz bands are cable TV systems, which use them in the Cable Television Relay Service (CARS) to convey video signals between points in their networks.

This includes the studio-to-headend link to connect an origination site with the site that transmits into the cable.

Cable TV companies are replacing many of these microwave CARS links with higher-capacity, two-way fiber optic links. Analysts believe that as much as half of this band could eventually be reallocated to other purposes.

TV stations use this spectrum for studio-transmitter links, and electronic news gathering. Cable companies use vacated CARS channels for electronic news gathering.

A variety of other fixed stations operate in this spectrum, some evicted from the 12.2–12.7 GHz DBS band.

NASA's Deep Space Network will use this band and 16.6–17.1 GHz for communication in the Space Research Service with spacecraft beyond the Earth's orbit. The band also is allocated to space research for reception only at NASA's Goldstone, California facility.

## 12.75-13.25 GHz

Gov: US251

Non-Gov: FIXED. FIXED-SATELLITE (Earth-to-space). MOBILE. 792A US251 NG53 NG104 NG118

FCC: AUXILIARY BROADCASTING (74). CABLE TELEVISION RELAY (78). DOMESTIC PUBLIC FIXED (21). PRIVATE OPERATIONAL-FIXED MICROWAVE (94).

See 12.7–12.75 GHz for the uses of this band. The American Mobile Satellite Corp. uses 13–13.15 and 10.75–10.95 GHz to feed its satellite (see 1530–1535 MHz).

## 13.25-13.4 GHz

Gov: AERONAUTICAL RADIONAVIGATION. Space Research (Earth-to-space). 851

Non-Gov: AERONAUTICAL RADIONAVIGATION. Space Research (Earth-to-space). 851

FCC: AVIATION (87).

Airborne Doppler radar navigation aids use the 13.25–13.4 and 8.75–8.85 GHz bands.

This band also is allocated, on a secondary basis, to satellite uplinks in the Space Research Service.

This is one of many restricted bands in which the FCC's Part 15 rules permit unlicensed devices to emit only very low level emissions.

## 13.4-13.75 GHz

Gov: RADIOLOCATION. Standard Frequency and Time Signal-Satellite (Earth-to-space). Space Research. 713 US110 G59

Non-Gov: Radiolocation. Standard Frequency and Time Signal-Satellite (Earth-to-space). Space Research. 713 US110

FCC: Private land mobile (90).

Mobile radars operate in this and the 13.75–14 GHz band worldwide. In the U.S., Navy and Coast Guard radars are the main uses. The Navy's Geosat program will operate satellite radar altimeters in this band. Geosat transmits oceanography data to Navy ships, government weather and space scientists and university researchers.

Satellite altimeters transmit radio energy and measure the reflected energy to obtain wave height, a measure of surface wind speed. Satellite altimetry is the only known way to precisely measure sea surface topography. The technique reveals changes in sea level prompted by such factors as warming of the global climate.

This band is available for police radars but reportedly is little used for this purpose. Future low-detectability radars are expected to use the band.

There is no U.S. use of this band, or its companion 20.2–21.2 GHz, for satellite frequency and time communications.

## 13.75-14 GHz

Gov: RADIOLOCATION. Standard Frequency and Time Signal-Satellite (Earth-to-space). Space Research 713 855A 855B 855C US110 USxxx G59

Non-Gov: (Proposed:) FIXED-SATELLITE (Earth-to-space). Radiolocation. Standard Frequency and Time Signal-Satellite (Earth-to-space). Space Research 713 855A 855B 855C US110 USxxx

FCC: (Proposed:) SATELLITE COMMUNICATION (25). Private land mobile (90).

Mobile radars operate in this and the 13.4–13.75 GHz band worldwide.

NASA's Tracking and Data Relay Satellite System uses this band's secondary Space Research Service allocation (see 1990–2110 MHz). The Space Shuttle can use a rendezvous radar for satellite retrieval in this band.

Other NASA missions use the band for active sensors. These transmit signals to a target and receive reflected signals, which are compared with transmitted signals. Active sensors include scatterometers, which measure wind velocity; altimeters; and precipitation radars.

A radar satellite using this band, part of the U.S.-Japan Tropical Rainfall Measuring Mission (TRMM), will be launched in 1997. TRMM will investigate rainfall over the tropics and subtropics. International Footnote 855C is intended to protect the TRMM radar from the Fixed Satellite Service (FSS) for a limited period.

The 1992 World Administrative Radio Conference allocated this band worldwide for FSS uplinks on a co-primary basis with radiolocation. The FCC proposed to accommodate the allocation in view of high FSS demand.

## 14-14.2 GHz

Gov: RADIONAVIGATION. Space Research. US287 US292

Non-Gov: FIXED-SATELLITE (Earth-to-space). RADIONAVIGATION. Space Research. US287 US292

FCC: Aviation (87). MARITIME (80). SATELLITE COMMUNICATION (25).

The Ku-band satellite uplink is 14–14.5 GHz (see 11.7–12.2 GHz). The band also is available to maritime radars, non-government land mobile-satellite uplinks and to airborne radionavigation on a secondary basis.

Satellite scatterometers use this spectrum to observe ocean winds by reflecting microwave pulses across broad regions of the Earth's surface.

The Space Station will use 14–14.3 and 14.5–14.89 GHz for communications with spacecraft in its near vicinity.

## *14.2-14.3 GHz*

Gov: US287

Non-Gov: FIXED-SATELLITE (Earth-to-space) US287

FCC: Satellite communications (25).

This is part of the Ku satellite uplink band (see 11.7–12.2 GHz). The 14.2–14.4 GHz spectrum is available for TV uses such as electronic news gathering.

## *14.3-14.4 GHz*

Gov: Fixed-satellite (Earth-to-space). Mobile except aeronautical mobile. US287

Non-Gov: FIXED-SATELLITE (Earth-to-space). US287

FCC: Satellite communication (25). Domestic public fixed (21).

This is part of the 14.2–14.4 GHz common carrier microwave band used for television services.

## *14.4-14.5 GHz*

Gov: Fixed. Mobile. 862 US203 US287

Non-Gov: FIXED-SATELLITE (Earth-to-space). 862 US203 US287

FCC: SATELLITE COMMUNICATION (25).

The Federal Aviation Administration uses 14.4–15.35 GHz for television microwave links that convey radar and video imagery between air facilities.

Uplinks in the Fixed Satellite Service should be the major activity in this band in the future.

Radio astronomers use this band and 4.83 GHz to detect formaldehyde, a component of the molecular structure of galaxies. The 14.47–14.5 GHz segment is one of many restricted bands in which the FCC's Part 15 rules permit unlicensed devices to emit only very low level emissions.

## 14.5-14.7145 GHz

Gov: FIXED. Mobile. Space Research.

Non-Gov: None

The Defense Department, NASA and the Federal Aviation Administration use this band. Military uses include electronic warfare, radar microwave links and cross-section measurement and threat simulation.

NASA's Tracking and Data Relay Satellite System has uplinks at 14.5–15.35 GHz (see 1990–2110 MHz). This is part of the FAA television microwave link band that relays video between aviation facilities.

## 14.7145-15.1365 GHz

Gov: MOBILE. Fixed. Space Research. US310

Non-Gov: US310

See 14.5–14.7145 GHz for the uses of this band.

## 15.1365-15.35 GHz

Gov: FIXED. Mobile. Space Research. 720 US211

Non-Gov: 720 US211

Air traffic control links, electronic warfare and government data and control links, radio astronomy and satellite-based rain and water vapor sensors use this band.

## 15.35-15.4 GHz

Gov: EARTH EXPLORATION-SATELLITE (passive). RADIO ASTRONOMY. SPACE RESEARCH (passive). US74 US246

Non-Gov: EARTH EXPLORATION-SATELLITE (passive). RADIO ASTRONOMY. SPACE RESEARCH (passive). US74 US246

No transmissions are permitted in this receive-only scientific band. It is one of several used to observe energy from quasars.

The 15.35–16.2 GHz spectrum is one of many restricted bands in which the FCC's Part 15 rules permit unlicensed devices to emit only very low level emissions.

## 15.4-15.7 GHz

Gov: AERONAUTICAL RADIONAVIGATION 733 797 US211 US260

Non-Gov: AERONAUTICAL RADIONAVIGATION 733 797 US211 US260

FCC: AVIATION (87).

This band was designated but not implemented for the Microwave Scanning Beam Landing System (MSBLS). MSBLS was to be part of the Microwave Landing System (MLS) which mainly operates in 5–5.25 GHz.

The 15.35–16.2 GHz spectrum is one of many restricted bands in which the FCC's Part 15 rules permit unlicensed devices to emit only very low level emissions.

## 15.7-16.6 GHz

Gov: RADIOLOCATION. US110 G59

Non-Gov: Radiolocation. US110

FCC: Private Land Mobile (90).

This is the most intensively used of the five bands from 15.7 to 17.7 GHz allocated to radiolocation.

Military uses include ground and airborne radars, telemetering and telecommand of aircraft, airborne weapons control, mortar location and combat surveillance.

This is an important part of the spectrum for airborne synthetic aperture radar (SAR), which generates higher resolution images than generated from conventional radar. Airborne SAR processes radar signals obtained along a flight path to create, in effect, a large antenna.

A SAR system provides its own "illumination"—the transmitted radio energy—and therefore can "see" through conditions such as night, clouds and dust that foil conventional imaging technology.

SAR has many applications in terrain mapping, geology and oceanography; treaty verification and weapons nonproliferation; reconnaissance, surveillance and targeting. SAR systems in various bands are flown onboard the Space Shuttle and satellites as well as on aircraft.

The Federal Aviation Administration uses the 15.7–16.2 GHz segment for Airport Surface Detection Equipment (ASDE). These are radars that monitor aircraft, vehicles and obstructions at ground facilities.

ASDE helps ensure that aircraft are on their proper runways or taxiways, particularly when weather restricts visibility. Latest versions of ASDE permit controllers to view maps of the airport that can be zoomed and rotated. Older ASDE systems operate at 23.6–24.4 GHz.

The 15.35–16.2 GHz spectrum is one of many restricted bands in which the FCC's Part 15 rules permit unlicensed devices to emit only very low level emissions.

## 16.6–17.1 GHz

Gov: RADIOLOCATION. Space Research (deep space). (Earth-to-space). US110 G59

Non-Gov: Radiolocation. US110

See 15.7–16.6 GHz for the uses of this band.

NASA's Deep Space Network expects to use this band and 12.75–13.25 GHz for future communication with spacecraft beyond Earth's orbit.

## 17.1–17.2 GHz

Gov: RADIOLOCATION. US110 G59

Non-Gov: Radiolocation. US110

See 15.7–16.6 GHz for the uses of this band.

## 17.2–17.3 GHz

Gov: RADIOLOCATION. Earth Exploration-Satellite (Active). Space Research (Active). US110 G59

Non-Gov: Radiolocation. Earth Exploration-Satellite (Active). Space Research (Active). US110

FCC: Private Land Mobile (90).

See 15.7–16.6 GHz for the uses of this band.

Besides the government radars in this band, active sensors onboard remote sensing satellites use the band to measure vegetation and snow, a use of the Earth Exploration-Satellite Service.

Active sensors transmit signals and then measure the signals reflected from the sensed object. All active sensor frequencies are in radiolocation bands.

## 17.3-17.7 GHz

Gov: Radiolocation. US259 US271 G59

Non-Gov: FIXED-SATELLITE (Earth-to-space). US259 US271 NG140

See 15.7–16.6 GHz for the uses of this band.

Fixed Satellite Service uplinks use this band, as do uplinks for Direct Broadcast Satellites in 17.3–17.8 GHz (DBS, see 12.2–12.7 GHz).

The 17.3–17.8 GHz band is supposed to become available under international agreement for High-Definition TV DBS downlinking in ITU Region 2 (the Americas) in 2007; the uplink will be at 24.75–25.25 GHz. The rest of the world will downlink HDTV DBS at 21.4–22.0 GHz.

## 17.7-17.8 GHz

Gov: US271

Non-Gov: FIXED. FIXED-SATELLITE (space-to-Earth) (Earth-to-space). MOBILE. US271 NG140 NG144

FCC: AUXILIARY BROADCASTING (74). CABLE TELEVISION RELAY (78). DOMESTIC PUBLIC FIXED (21). PRIVATE OPERATIONAL-FIXED MICROWAVE (94).

Uses of the 17.7–19.7 GHz general-purpose fixed spectrum include Broadcast Auxiliary, common carrier, cable TV and private radio users for data communications.

The Ka-band consists of satellite downlinks in 17.7–20.2 GHz, paired with uplinks at 27.5–30 GHz.

Competing with Teledesic, a system in the Non-Geostationary Fixed Satellite Service (see 17.8–18.6 GHz), are many proposed broadband Geostationary Fixed Satellite Service (GSO/FSS) ventures that plan to use the Ka-band. They will provide intriguing high-speed, Internet-like services, beginning around year 2000, at startup costs in the billions of dollars. Their designated downlink segments are 19.7–20.2 and 17.7–18.8 GHz, paired with uplinks at 29.5–30 and 28.35–28.6 GHz.

Some of these satellites also employ intersatellite links in 59–64 GHz. They typically use 3.7–4.2 and 5.925–6.425 GHz (C-band) for telemetry and control during transfer to their eventual orbit positions.

The GSO/FSS should have tremendous social and economic impact in the next century.

They should not be confused with the Non-Geostationary Mobile Satellite Service, as represented by Iridium (see 1610–1610.6 MHz): the Big LEO systems for portable telephones. The GSO/FSS systems eventually may connect to portable, even miniature, user terminals and compete with Big LEOs for telephone customers.

AT&T proposed VoiceSpan, a 12-satellite system designed to support more than 12 million subscribers. "VoiceSpan will span the gap between an advanced multimedia infrastructure and more traditional voice-oriented systems," the company said. "Services available through the VoiceSpan system will include electronic messaging and mailboxes, information and database access, multimedia bridging, software distribution and updates, financial transactions, and voice communications."

Hughes Communications Galaxy proposed the $5.1 billion Galaxy/Spaceway, a combined broadband GSO/FSS and direct broadcast satellite (DBS) system of 20 satellites. Indicative of the size and complexity of this proposal, the FCC's filing fee alone for the Galaxy/Spaceway license application approached $1 million. "Galaxy/Spaceway customers can use a single ultra-small satellite antenna for both high speed access to the information superhighway and for direct-to-home video reception," Hughes said.

Lockheed Martin proposed a $4 billion international system of nine satellites called Astrolink. "The Astrolink system will serve a totally different market from 'Big' and 'Little' LEOs which primarily seek to provide narrowband voice and data service to mobile handheld terminals," the company said.

Astrolink will "target a new generation of users requiring low cost and wideband communication from both fixed and temporary locations." Projected markets include distribution of software and music; satellite news gathering; global Internet service and wireless backbone for worldwide financial networks.

Motorola subsidiary Comm, Inc. proposed a $2.3 billion, four-satellite system code-named Millenium. "Motorola intends to transplant the concept of 'any person, anytime, anywhere" communications—the goal of the Iridium system in the area of the Mobile Satellite Service—to broadband communications in the Fixed Satellite Service," the company said. Millenium will offer "bandwidth on demand" to customers in North, Central and South America using asynchronous transfer mode (ATM) at data rates ranging up to 51.84 Mbps.

EchoStar's $340 million, two-satellite domestic system will provide services that include "video telephony, videoconferencing, voice communications, high-speed personal computer access to on-line information and entertainment services, medical and technical tele-imaging," and messaging services to land and maritime mobile users. EchoStar operates the DISH Network direct broadcast satellite service (see 12.2–12.7 GHz).

KaSTAR, which bills itself as "America's Interactive Satellite System," proposes to launch two "Ladybug" satellites at a cost of $605 million. KaSTAR will provide ATM-based "low cost, ubiquitous high rate digital data, video telephony service" which will drive the merging of "computers, phones, video and television into one interactive and seamless whole."

Other applicants for Ka-band GSO/FSS satellite systems include GE American Communications, which proposed the GE*Star system of nine satellites, to make available "ubiquitous and affordable broadband digital service to virtually the entire globe"; Morningstar, which proposed a $847 million four-satellite international system; Netsat 28, which proposed a $250 million domestic satellite for "ubiquitous nationwide access" to broadband services; Orion Asia Pacific, Orion Atlantic and Orion Network Systems, which proposed $757 million worth of international and domestic satellites to provide video, audio and multimedia services; Loral Aerospace Holdings, which proposed CyberStar, a $1.05 billion "global wireless interactive multimedia transmission system" with three satellites; PanAmSat, which proposed two satellites at $409 million for the U.S., Latin America, Europe and Western Africa; and VisionStar, a $207.5 million domestic satellite for video programming and conferencing services.

DBS uplinks have access to the 17.3–17.8 and 17.8–18.4 GHz spectrum (see 12.2–12.7 GHz). Big LEO mobile satellites will use feeder links in the Ka-band Fixed Satellite Service allocations (see 1610–1610.6 MHz).

The 17.7–21.4 GHz spectrum is one of many restricted bands in which the FCC's Part 15 rules permit unlicensed devices to emit only very low level emissions.

---

## 17.8-18.6 GHz

Gov: 870 US334 G117

Non-Gov: FIXED. FIXED-SATELLITE (space-to-Earth). MOBILE. 870 US334 NG144

FCC: AUXILIARY BROADCASTING (74). CABLE TELEVISION RELAY (78). DOMESTIC PUBLIC FIXED (21). PRIVATE OPERATIONAL-FIXED MICROWAVE (94).

Uses of the 17.7–19.7 GHz general-purpose fixed spectrum include Broadcast Auxiliary, common carrier, cable TV and private radio users for data communications.

This band is part of the Ka-band satellite allocations at 17.7–20.2 GHz for downlinks, paired with uplinks at 27.5–30 GHz (see 17.7–17.8 GHz).

International footnote 870 provides that 18.1–18.3 GHz is allocated to uplinks in the meteorological-satellite service on a co-primary basis for geostationary satellites.

Recently-added footnote US334 permits government fixed satellite downlinks to operate in 17.8–20.2 GHz on a primary basis, but only if they are outside a specified orbital location and coordinated in frequency with non-government systems.

The U.S. conducts, or plans to conduct, certain defense satellite communications in this band. The National Telecommunications and Information Administration in July 1995 announced that this allocation involved specific sensitive "urgent" national security interests of the United States, which it avoided describing in detail.

Accordingly, the FCC revised footnote G117 to reflect that government satellite systems in this portion of the spectrum are limited to military systems.

Microsoft's Bill Gates and communications magnate Craig McCaw are investors in a $9.55 billion broadband satellite project. Their Teledesic system will use downlinks at 18.8–19.3 GHz with uplinks at 28.6–29.1 GHz.

Calling Communications invented the system and announced it in 1993 to little press attention. Gates and McCaw acquired Calling Communications and renamed it Teledesic Corp. Teledesic filed a formal license application at the FCC on March 21, 1994, this time generating a blizzard of news stories.

The system can only be described as massive. Teledesic proposes to place into low-Earth orbit 840 satellites plus up to 84 on-orbit spares. The satellites resemble upside-down flowers of eight petals. Each jointed petal has three panels. Each panel is a transceiver with phased-array antenna.

The company described its service as a "global Internet," offering voice channels, broadband channels supporting videoconferencing and interactive multimedia, and real-time two-way data services. Supported data speeds range from 16 kbps to 1.244 Gbps. It could transmit video with "VCR quality." Each satellite is a node in a packet network, with communication links to eight adjacent satellites.

Teledesic is supposed to be for fixed, not mobile, use, as it is classified as a Non-Geostationary Fixed Satellite Service (NGSO/FSS). We expect users to ignore such official distinctions, using Teledesic radios while moving or stationary. Teledesic expects to begin launching in the year 2000, with regular service beginning in 2001.

Competing with Teledesic will be several Geostationary Fixed Satellite Service ventures (see 17.7–17.8 GHz).

The 17.7–21.4 GHz spectrum is one of many restricted bands in which the FCC's Part 15 rules permit unlicensed devices to emit only very low level emissions.

## 18.6-18.8 GHz

Gov: EARTH EXPLORATION-SATELLITE (passive). SPACE RESEARCH (passive). US254 US255 US334 G117

Non-Gov: EARTH EXPLORATION-SATELLITE (passive). FIXED. FIXED-SATELLITE (space-to-Earth). MOBILE except aeronautical mobile. SPACE RESEARCH (passive). US254 US255 US334 NG144

FCC: CABLE TELEVISION RELAY (78). DOMESTIC PUBLIC FIXED (21). PRIVATE OPERATIONAL-FIXED MICROWAVE (94).

This is part of the general-purpose fixed spectrum at 17.7–19.7 GHz and the Ka-band satellite allocations at

17.7–20.2 GHz for downlinks, paired with uplinks at 27.5–30 GHz (see 17.7–17.8 GHz).

Earth exploration satellites use the 18.6–18.8 GHz segment to measure rainfall.

The 17.7–21.4 GHz spectrum is one of many restricted bands in which the FCC's Part 15 rules permit unlicensed devices to emit only very low level emissions.

## 18.8-19.7 GHz

Gov: US334 G117

Non-Gov: FIXED. FIXED-SATELLITE (space-to-Earth). MOBILE. US334 NG144

FCC: AUXILIARY BROADCASTING (74). CABLE TELEVISION RELAY (78). DOMESTIC PUBLIC FIXED (21). PRIVATE OPERATIONAL-FIXED MICROWAVE (94).

This band is part of the 17.7–19.7 GHz general-purpose fixed spectrum. The band also is part of the Ka-band satellite allocations at 17.7–20.2 GHz for downlinks, paired with uplinks at 27.5–30 GHz (see 17.7–17.8 GHz).

The 19.3–19.7 GHz segment is designated for Big LEO downlink feeder links, paired with 28.6–29.1 GHz for feeder uplinks (see 1610–1610.6 MHz).

Parts of this spectrum are available to Digital Termination Systems (DTS), point-to-multipoint radio networks operating in the Private Operational Fixed Service. Common-carrier, fee-for-service operations using DTS are licensed under the Digital Electronic Message Service (DEMS, see 10.55–10.6 GHz).

Motorola petitioned the FCC and received changes to the DTS rules. These changes permit Motorola to market Altair, a line of low-power local area network radios at 18 and 19 GHz. Altair provides wireless data communications indoors and between buildings.

Altair is often mistakenly believed to be an unlicensed device. In fact, Motorola holds the license for each area where customers use Altair radios. The company obtained hundreds of licenses across the U.S. for this purpose.

Each license covers a 17.5 mile radius of a reference point. The licensee may operate multiple low-power point-to-multipoint transmitters in this band anywhere within the radius. Because of special FCC action, the federal government can be a customer of these LAN systems.

The 17.7–21.4 GHz spectrum is one of many restricted bands in which the FCC's Part 15 rules permit unlicensed devices to emit only very low level emissions.

## 19.7-20.1 GHz

Gov: US334 G117

Non-Gov: FIXED-SATELLITE (space-to-Earth). MOBILE-SATELLITE (space-to-Earth). 873A 873B 873C 873D 873E US334

This band is part of the 17.7–19.7 GHz general-purpose fixed spectrum. The band also is part of the Ka-band satellite allocations at 17.7–20.2 GHz for downlinks, paired with uplinks at 27.5–30 GHz (see 17.7–17.8 GHz).

NASA launched the Advanced Communications Technology Satellite (ACTS) in 1993. ACTS uses an onboard baseband processor, routing and switching traffic between source and destination earth stations.

Services developing from ACTS include satellite supercomputer access, satellite-based cellular and personal communications networks, narrowband mobile radio, portable maritime communications, video programming to passenger trains, and intelligent transportation systems.

The 17.7–21.4 GHz spectrum is one of many restricted bands in which the FCC's Part 15 rules permit unlicensed devices to emit only very low level emissions.

## 20.1-20.2 GHz

Gov: US334 G117

Non-Gov: FIXED-SATELLITE (space-to-Earth). MOBILE-SATELLITE (space-to-Earth). 873A 873B 873C 873D US334

See 19.7–20.1 GHz for the uses of this band.

The 17.7–21.4 GHz spectrum is one of many restricted bands in which the FCC's Part 15 rules permit unlicensed devices to emit only very low level emissions.

## 20.2-21.2 GHz

Gov: FIXED-SATELLITE (space-to-Earth). MOBILE-SATELLITE (space-to-Earth). Standard Frequency and Time Signal-Satellite (space-to-Earth). G117

Non-Gov: Standard Frequency and Time Signal-Satellite (space-to-Earth).

The Pentagon's Milstar has a downlink in this band, and an uplink at 43.5–45.5 GHz. Milstar is the controversial Military, Strategic and Tactical Relay satellite system to provide secure battlefield communications.

The six-satellite Milstar is the most expensive satellite system ever. Expenditures on Milstar are pegged at $17 billion through 2002. Controlled from Falcon Air Force Base in Colorado Springs, Colorado, Milstar provides communications for users in the field, in missile silos, submarines, bombers and special trucks that travel the highways during periods of high military alert.

Unlike other systems of particular branches of the armed forces, Milstar terminals are interoperable across all military services. Features such as EHF operation, time and frequency permutation and frequency hopping keep Milstar robust against enemy jamming or propagation disturbances (scintillation) from nuclear detonations.

EHF allows physically small, high-gain antennas, permitting mobility and low transmit power. The beamwidth of a Milstar EHF uplink is narrow, making it more difficult for an opponent to place interceptor aircraft in the beam. Milstar incorporates a lower-frequency UHF service (350 MHz uplink, 250 MHz downlink) compatible with older military satellite systems.

The satellites contain on-board message switching and 60 GHz crosslinks between satellites. The satellites themselves route messages, through crosslinks if necessary, to the appropriate downlink to make a connection. This permits communications without the need for ground relays or facilities on foreign soil.

Huge cost overruns, technical setbacks and whistleblowers' charges eventually brought the ostensibly classified program undesired scrutiny. An Air Force colonel told Congressional committees his views about Milstar cost him his job evaluating the space budget. He described Milstar as a costly monument to Cold War technology that should be terminated. America's Defense Monitor called Milstar "another expensive weapon system made obsolete by the collapse of Communism."

Officially, the Defense Department claims that nuclear warfighting capability was removed from Milstar and replaced with more conventional components.

In 1993, a Milstar power amplifier fell off a truck on its way to McClellan Air Force Base in California. The amplifier ended up on the unclaimed freight market. Not knowing exactly what the device was, Roger Spillman of Cooleemee, N.C. bought the $363,735 amplifier from a salvage dealer for $75.00. He had hoped to use it with a radio in his truck.

The Air Force seized the unit after a friend of Spillman's telephoned Raytheon, the manufacturer, to ask for an owner's manual. The national press and TV widely reported the story, launching Milstar into the public eye.

The first Milstar Development Flight Satellite, DFS-1, was placed in orbit February 7, 1994. Another was launched in 1995. Beginning in 1999, additional launches will deploy second-generation Milstar Block II spacecraft that will replace both Block I satellites. Each satellite has a design life of 10 years. Milstar's successor, the Advanced EHF Satellite System, is now being planned.

There reportedly is no U.S. use of this band yet, or its companion band 13.4–14 GHz, for two-way satellite frequency and time communications.

The 17.7–21.4 GHz spectrum is one of many restricted bands in which the FCC's Part 15 rules permit unlicensed devices to emit only very low level emissions.

## 21.2-21.4 GHz

Gov: EARTH EXPLORATION-SATELLITE (passive). FIXED. MOBILE. SPACE RESEARCH (passive). US263

Non-Gov: EARTH EXPLORATION-SATELLITE (passive). FIXED. MOBILE. SPACE RESEARCH (passive). US263

FCC: DOMESTIC PUBLIC FIXED (21). PRIVATE OPERATIONAL-FIXED MICROWAVE (94).

The 21.2–23.6 GHz spectrum is for general-purpose fixed uses such as phone bypass and links for aviation, TV stations, PBX telephone systems, computer networks, security cameras and video teleconferencing.

The band also is shared with spacecraft in the Earth Exploration-Satellite Service, such as Geosat (see 13.4–13.75 GHz), for measurement of the Earth's water vapor.

The 17.7–21.4 GHz spectrum is one of many restricted bands in which the FCC's Part 15 rules permit unlicensed devices to emit only very low level emissions.

## 21.4-22 GHz

Gov: FIXED. MOBILE.

Non-Gov: FIXED. MOBILE.

FCC: DOMESTIC PUBLIC FIXED (21). PRIVATE OPERATIONAL-FIXED MICROWAVE (94).

See 21.2–21.4 GHz for the uses of this band.

## 22-22.21 GHz

Gov: FIXED. MOBILE except aeronautical mobile. 874

Non-Gov: FIXED. MOBILE except aeronautical mobile. 874

FCC: DOMESTIC PUBLIC MOBILE (22). PRIVATE OPERATIONAL-FIXED MICROWAVE (94).

See 21.2–21.4 GHz for the uses of this band.

The 22.01–23.12 GHz spectrum is one of many restricted bands in which the FCC's Part 15 rules permit unlicensed devices to emit only very low level emissions.

## 22.21-22.5 GHz

Gov: EARTH EXPLORATION-SATELLITE (passive). FIXED. MOBILE except aeronautical mobile. RADIO ASTRONOMY. SPACE RESEARCH (passive). 875 US263

Non-Gov: EARTH EXPLORATION-SATELLITE (passive). FIXED. MOBILE except aeronautical mobile. RADIO ASTRONOMY. SPACE RESEARCH (passive). 875 US263

FCC: DOMESTIC PUBLIC FIXED (21). PRIVATE OPERATIONAL-FIXED MICROWAVE (94).

Atmospheric gases emit and absorb energy at discrete resonance frequencies. Molecules of water resonate at 22.235 GHz. Measurements around this frequency indicate the amount of water in an atmosphere. Radio astronomers use this data to measure distances to far galaxies. The band also is used for Very Long Baseline Interferometry by connected radio telescopes.

These passive (receive-only) uses are not protected against interference from active (transmitting) uses as part of the 21.2–23.6 GHz fixed spectrum (see 21.2–21.4 GHz).

The 22.01–23.12 GHz spectrum, however, is one of many restricted bands in which the FCC's Part 15 rules permit unlicensed devices to emit only very low level emissions.

## 22.5-22.55 GHz

Gov: FIXED. MOBILE. US211

Non-Gov: FIXED. MOBILE. US211

FCC: DOMESTIC PUBLIC FIXED (21). PRIVATE OPERATIONAL-FIXED MICROWAVE (94).

See 21.2–21.4 GHz for the uses of this band.

The 22.01–23.12 GHz spectrum is one of many restricted bands in which the FCC's Part 15 rules permit unlicensed devices to emit only very low level emissions.

## 22.55-23 GHz

Gov: FIXED. INTER-SATELLITE. MOBILE. 879 US278

Non-Gov: FIXED. INTER-SATELLITE. MOBILE. 879 US278

FCC: DOMESTIC PUBLIC FIXED (21). PRIVATE OPERATIONAL-FIXED MICROWAVE (94). SATELLITE COMMUNICATIONS (25).

This is part of the 21.2–23.6 GHz fixed microwave band (see 21.2–21.4 GHz).

It also is one of the bands allocated for intersatellite (satellite-to-satellite) links in both non-geostationary and geostationary systems. To be compatible with planned European and Japanese relay satellite systems, NASA targets 22.55–23.55 and 25.25–27.5 GHz for future Tracking and Data Relay Satellites (see 1990–2110 MHz).

The 22.01–23.12 GHz spectrum is one of many restricted bands in which the FCC's Part 15 rules permit unlicensed devices to emit only very low level emissions.

## 23-23.55 GHz

Gov: FIXED. INTER-SATELLITE. MOBILE. 879 US278

Non-Gov: FIXED. INTER-SATELLITE. MOBILE. 879 US278

FCC: DOMESTIC PUBLIC FIXED (21). PRIVATE OPERATIONAL-FIXED MICROWAVE (94).

See 21.2–21.4 GHz for the uses of this band.

The 22.01–23.12 GHz spectrum is one of many restricted bands in which the FCC's Part 15 rules permit unlicensed devices to emit only very low level emissions.

## 23.55-23.6 GHz

Gov: FIXED. MOBILE.

Non-Gov: FIXED. MOBILE.

FCC: DOMESTIC PUBLIC FIXED (21). PRIVATE OPERATIONAL-FIXED MICROWAVE (94).

See 21.2–21.4 GHz for the uses of this band.

## 23.6-24 GHz

Gov: EARTH EXPLORATION-SATELLITE (passive). RADIO ASTRONOMY. SPACE RESEARCH (passive). US74 US246

Non-Gov: EARTH EXPLORATION-SATELLITE (passive). RADIO ASTRONOMY. SPACE RESEARCH (passive). US74 US246

No transmissions are permitted in this scientific band. Molecular lines in this band indicate ammonia in the interstellar medium. Satellite sensors use the band to measure water vapor. This is one of many restricted bands in which the FCC's Part 15 rules permit unlicensed devices to emit only very low level emissions.

## 24-24.05 GHz

Gov: 881 US211

Non-Gov: AMATEUR. AMATEUR-SATELLITE. 881 US211

FCC: AMATEUR (97).

This portion of the spectrum contains Amateur Radio experiments, active satellite-based sensors that measure vegetation and snow, unlicensed communications devices and Industrial, Scientific and Medical equipment.

## 24.05-24.25 GHz

Gov: RADIOLOCATION. Earth Exploration-Satellite (Active). 881 US110 G59

Non-Gov: Amateur. Radiolocation. Earth Exploration-Satellite (Active). 881 US110

FCC: Amateur (97). Private Land Mobile (90).

K-band police radars operate at 24.15 GHz. Consumer electronics manufacturers are developing a new Safety Warning Transmitter System (SWTS) at 24.1 GHz. SWTS units are capable of measuring the speed of nearby traffic and sending warning messages to consumer radar detectors.

Another application would use a transmitter mounted on an emergency vehicle. The transmitter would illuminate the radar detector from behind, alerting the driver to an approaching emergency vehicle.

The Side Detection System in this band alerts the driver if an object with radar cross section of one square meter or greater is present in the vehicle's blind zone.

## 24.25-24.45 GHz

Gov: RADIONAVIGATION.

Non-Gov: RADIONAVIGATION.

FCC: AVIATION (87).

## 24.45-24.65 GHz

Gov: INTER-SATELLITE. RADIONAVIGATION. 882E

Non-Gov: INTER-SATELLITE. RADIONAVIGATION. 882E

FCC: SATELLITE COMMUNICATION (25).

## 24.65-24.75 GHz

Gov: INTER-SATELLITE. RADIOLOCATION-SATELLITE (Earth-to-space).

Non-Gov: INTER-SATELLITE. RADIOLOCATION-SATELLITE (Earth-to-space).

FCC: SATELLITE COMMUNICATION (25).

## 24.75-25.25 GHz

Gov: RADIONAVIGATION.

Non-Gov: RADIONAVIGATION.

FCC: AVIATION (87).

## 25.25-27 GHz

Gov: FIXED. MOBILE. Earth Exploration-Satellite (space-to-space). Standard Frequency and Time Signal-Satellite (Earth-to-space).

Non-Gov: Earth Exploration-Satellite (space-to-space). Standard Frequency and Time Signal-Satellite (Earth-to-space).

International agreements prescribe 25.25–27.5 GHz for links between satellites in the Space Research Service and

Earth Exploration-Satellite Service, and for data from industrial and medical activities in space.

The Air Force is developing the Wireless Broadband System (WBS) in 25.25–27.5 GHz. WBS is a one- and two-way voice, video and data distribution network for use at theatre air bases.

There reportedly is no U.S. use of this band yet, or its companion band 30–31.3 GHz, for two-way satellite frequency and time communications.

NASA targets the 25.25–27.5 and 22.55–23.55 GHz bands for future Tracking and Data Relay Satellites (see 1990–2110 MHz).

## *27-27.5 GHz*

Gov: FIXED. MOBILE. Earth Exploration-Satellite (space-to-space).

Non-Gov: Earth Exploration-Satellite (space-to-space).

This is part of the spectrum used for the Air Force Wireless Broadband System (see 25.25–27 GHz).

## *27.5-29.5 GHz*

Gov: None

Non-Gov: FIXED. FIXED-SATELLITE (Earth-to-space). MOBILE.

FCC: DOMESTIC PUBLIC FIXED (21).

The FCC divided this portion of the Ka-band into segments to accommodate the Local Multipoint Distribution Service (LMDS), a "cellular cable" broadband fixed service, and mobile and fixed satellite services.

LMDS competes with cable TV, wireless cable (see 2150–2160 and 2596–2644 MHz), VHF and UHF TV, direct broadcast satellites (see 12.2–12.7 GHz), and is expected to compete with local telephone service.

LMDS uses transmitters at user sites, as well as cell sites, for two-way communication. The FCC predicts that video programming will be the main LMDS offering, but it will permit “any video or telecommunications use”.

The LMDS license is permissive. The FCC said licensees “would not be restricted to specific bandwidth, emission characteristics, etc. and could change the traffic mix within the frequency assignment to meet the requirements of the individual community served by a cell or multiple cells.”

CellularVision, successor to original LMDS developers, has been the only operator licensed to provide LMDS in the U.S. on a revenue basis. The company provides wireless video service in New York.

The New York system is an unusual case. The LMDS developers persuaded the FCC to grant them a regular LMDS license before the FCC established LMDS, allocated spectrum to it, or adopted regulations for it.

Other companies have experimental, non-revenue licenses for LMDS in various U.S. cities, and LMDS operations are developing in other countries including Brazil, Canada, Venezuela and Guatemala. LMDS license auctions in the U.S. are a certainty.

LMDS has an 850 MHz-wide primary segment in 27.5–28.35 GHz and a companion 150 MHz-wide segment in 29.1–29.25 GHz, shared with Big LEO feeder links (see 1610–1610.6 MHz).

The FCC proposed to issue an additional 300 MHz of spectrum to LMDS by reallocating 31.0–31.3 GHz, probably stifling unique applications there. Incumbent 31 GHz users vehemently reject reallocation, especially since the FCC may not renew their licenses.

Systems in the Geostationary Fixed Satellite Service (GSO/FSS) have primary uplink allocations in 28.35–28.6 GHz and 29.5–30 GHz, and a co-primary allocation shared with Big LEO feeder uplinks in 29.25–29.5 GHz (see 17.7–17.8 GHz for more information on GSO/FSS).

The Non-Geostationary Fixed Satellite Service (NGSO/FSS) has a primary uplink segment at 28.6–29.1 GHz and might use 28.35–28.6 and 29.5–30 GHz on a secondary basis. The only U.S. NGSO/FSS applicant at this writing is Teledesic (see 17.8–18.6 GHz).

## 29.5-29.9 GHz

Gov: 882

Non-Gov: FIXED-SATELLITE (Earth-to-space). Mobile-Satellite (Earth-to-space). 882

The 29.5–30 GHz portion of the 27.5–30 GHz Ka-band uplink spectrum is devoted to Geostationary Fixed Satellite Service (GSO/FSS) systems on a primary basis and Non-Geostationary Fixed Satellite Service (NGSO/FSS) on a secondary basis (see 27.5–29.5 GHz).

## 29.9-30 GHz

Gov: 882

Non-Gov: FIXED-SATELLITE (Earth-to-space). Mobile-Satellite (Earth-to-space). 882

See 29.5–29.9 GHz for information on this band.

## 30-31 GHz

Gov: FIXED-SATELLITE (Earth-to-space). MOBILE-SATELLITE (Earth-to-space). Standard Frequency and Time Signal-Satellite (space-to-Earth). G117

Non-Gov: Standard Frequency and Time Signal-Satellite (space-to-Earth).

"If bandwidth is real estate," writes Andrew Clegg of *Millimeter Wave Report*, "then nowhere is there more open space than the millimeter wave band. Over 90% of the allocated radio spectrum falls above 30 GHz."

The 30–31 GHz band is one of several allocated to military fixed and mobile satellites. There is no U.S. use of this band yet, or its companion band 25.25–27 GHz, for two-way satellite frequency and time communications.

Radio astronomers are planning an immense Millimeter Array (MMA) radio telescope installation to operate at frequencies from 30 GHz to more than 600 GHz.

The MMA will use 40 transportable, parabolic dish antennas with carefully manufactured reflective surfaces. The antennas can be placed in a ring or in other formations depending on the particular observations sought.

To be built in a special high-altitude location in Chile or Hawaii, the MMA will help scientists understand the age of the universe; image distant galaxies and pre-planetary structures; measure the properties of subsurface layers of asteroids; and probe the energetics of the outflows from thousands of stars, among its many possibilities.

Another facility, the joint U.S.-Mexican Large Millimeter Wave Telescope (LMT), will be located on a mountain in Mexico. The $50 million, 50-meter telescope will be able to point to a cosmic source with an accuracy of 1/1000th of a degree—the size of a dime at a distance of 10 miles.

"Millimeter-wave astronomy enables scientists to see through much of the dust and haze that obscures stellar activity," says astronomer Andrew Harris of the University of Massachusetts, Amherst. "The LMT project could help shed light on what we don't yet understand about how galaxies, stars and planets begin."

# 31-31.3 GHz

Gov: Standard Frequency and Time Signal-Satellite (space-to-Earth). 886 US211

Non-Gov: FIXED. MOBILE. Standard Frequency and Time Signal-Satellite (space-to-Earth). 886 US211

FCC: AUXILIARY BROADCASTING (74). DOMESTIC PUBLIC FIXED (21). CABLE TELEVISION RELAY (78). GENERAL MOBILE RADIO (95). PRIVATE OPERATIONAL-FIXED MICROWAVE (94).

This little-known band is becoming more popular for fixed service use. Sierra Digital Communications pioneered equipment in this band for traffic signal control.

The conventional frequency coordination procedures and costs associated with other fixed service bands do not apply to 31–31.3 GHz. Systems in this band are marketed as a low cost alternative to leased telephone lines, fiber or coaxial cable or links in traditional microwave bands.

In addition to the commercial fixed services, in 1985 the agency made this band available to individuals for personal use by allocating it to the General Mobile Radio Service (see 462.5375–462.7375 MHz). The FCC later relaxed the antenna rules for this band to support mobile use.

No interference protection is afforded to fixed or mobile stations operating in this band. Use of this spectrum is not permitted for direct delivery of video programs to the general public or for multichannel cable TV distribution.

The FCC allocated this band to "satisfy various types of short range, fixed and mobile communications requirements. ...The proposal was based on suggestions from commenters...who stated that there is a need for a spectrum allocation having greatly reduced licensing and coordination requirements.

"The Commission is trying to find spectrum for those who do not fall within the traditional service categories, but yet have unfulfilled communication requirements. We continue to feel that this 300 megahertz at 31 GHz is an appropriate location for use by individuals. One of the driving forces behind this proceeding is the reduction or elimination of 'type of entity' barriers, except perhaps where some special conditions exist."

In 1996 the FCC placed existing and future 31 GHz uses in jeopardy by proposing to reallocate the band to the Local Multipoint Distribution Service (see 27.5–29.5 GHz).

The 31.2–31.8 GHz spectrum is one of many restricted bands in which the FCC's Part 15 rules permit unlicensed devices to emit only very low level emissions.

## 31.3-31.8 GHz

Gov: EARTH EXPLORATION-SATELLITE (passive). RADIO ASTRONOMY. SPACE RESEARCH (passive). US74 US246

Non-Gov: EARTH EXPLORATION-SATELLITE (passive). RADIO ASTRONOMY. SPACE RESEARCH (passive). US74 US246

No transmissions are permitted in this receive-only scientific band. Satellite-based sensors use it to measure cloud water, precipitation and changes in ice flows.

The 31.2–31.8 GHz spectrum is one of many restricted bands in which the FCC's Part 15 rules permit unlicensed devices to emit only very low level emissions.

## 31.8-32 GHz

Gov: RADIONAVIGATION. US69 US211 US262

Non-Gov: RADIONAVIGATION. US69 US211 US262

This band is available to airborne radionavigation and to downlinks in deep space communications, limited to NASA's Goldstone, California facility.

## 32-33 GHz

Gov: INTER-SATELLITE. RADIONAVIGATION. 893 US69 US262 US278

Non-Gov: INTER-SATELLITE. RADIONAVIGATION. 893 US69 US262 US278

The Cassini mission to Saturn will use links in the 32–33.4 GHz spectrum. In addition to the deep space and intersatellite uses of this band, the Space Station may use it to connect with non-U.S. satellites.

## 33-33.4 GHz

Gov: RADIONAVIGATION. US69

Non-Gov: RADIONAVIGATION. US69

See 32–33 GHz for information about this band.

## 33.4-36 GHz

Gov: RADIOLOCATION. 897 US110 US252 G34

Non-Gov: Radiolocation. 897 US110 US252

FCC: Private Land Mobile (90).

NASA uses this band for deep space uplinks. Newer, Ka-band police radars and satellite cloud and snow sensors also operate in this band.

## 36-37 GHz

Gov: EARTH EXPLORATION-SATELLITE (passive). FIXED. MOBILE. SPACE RESEARCH (passive). 898 US263

Non-Gov: EARTH EXPLORATION-SATELLITE (passive). FIXED. MOBILE. SPACE RESEARCH (passive). 898 US263

Passive sensors onboard spacecraft such as Geosat (see 13.4–13.75 GHz) use this band to measure rain rate, snow water content and ocean ice.

The 36.43–36.5 GHz segment is one of many restricted bands in which the FCC's Part 15 rules permit unlicensed devices to emit only very low level emissions.

## *37-38.6 GHz*

Gov: FIXED. MOBILE.

Non-Gov: FIXED. MOBILE.

FCC: DOMESTIC PUBLIC FIXED (21). PRIVATE OPERATIONAL-FIXED MICROWAVE (94).

The 37–40 GHz spectrum is of intense interest to Personal Communications Services (PCS) carriers (see 1850–1990 MHz) and cellular carriers (see 824–849 MHz) for interconnecting base stations within their systems.

At this writing there is no non-government use of 37–38.6 GHz and little government use of the band. The 38.6–40 GHz spectrum is in wide use and the subject of a huge number of license applications. Carriers use 38.6–40 GHz to connect business customers, long distance carriers and PCS carriers. TV Broadcast Auxiliary stations use 38.6–40 GHz to connect remote and transmission sites.

"The frequencies at 39 GHz, once hardly noticed, have now become highly desirable, largely due to the development of innovative technologies and services," according to FCC Commissioner Susan Ness.

The FCC proposed to make 37–38.6 GHz available for point-to-point fixed service use, and to essentially combine the band with adjacent spectrum to create a unified 37–40 GHz band. Licenses for channel blocks in this spectrum would be auctioned in 493 geographic divisions of the U.S. called Basic Trading Areas (BTAs).

The FCC also is considering allocating all or part of 37–40 GHz for point-to-multipoint systems (like wireless cable) and/or mobile use. These uses would be licensed, probably by auction, in the General Wireless Communications Service (see 4.66–4.685 GHz) or Licensed Millimeter Wave Service (see 40.5–42.5 GHz).

"We believe that the 37 GHz band has several similarities to GWCS," the FCC said, "such as relatively small geographic areas, no eligibility limitations, uniform channel blocks, and the potential for a variety of different uses. Although the 37 GHz band may be used for backhaul [between base stations and the switching office] and backbone [between switching offices] communications links for broadband PCS, our proposed rules do not require such use. Also, many more licenses will be available in the 37 GHz band than were available for broadband PCS."

The National Telecommunications and Information Administration asked that 37–38 GHz be allocated to Government Space Research Service downlinks. The FCC asked if such operations could share with fixed stations.

NTIA—which ordinarily is concerned with government spectrum use—also asked the FCC to allocate 37–38 GHz to Space Research Service downlinks, on a secondary basis, to non-government use. The FCC responded that it was unaware of any need for non-government entities to have access to this spectrum for space research purposes.

## *38.6-39.5 GHz*

Gov: US291

Non-Gov: FIXED. MOBILE. FIXED-SATELLITE (space-to-Earth). US291

FCC: DOMESTIC PUBLIC FIXED (21). PRIVATE OPERATIONAL-FIXED MICROWAVE (90). Auxiliary Broadcasting (74).

See 37–38.6 GHz for information about this band.

## 39.5-40 GHz

Gov: FIXED-SATELLITE (space-to-Earth). MOBILE-SATELLITE (space-to-Earth). US291 G117

Non-Gov: FIXED. FIXED-SATELLITE (space-to-Earth). MOBILE. MOBILE-SATELLITE (space-to-Earth). US291

FCC: DOMESTIC PUBLIC FIXED (21). PRIVATE OPERATIONAL-FIXED MICROWAVE (94). Auxiliary Broadcasting (74).

See 37–38.6 GHz for information about this band.

## 40-40.5 GHz

Gov: FIXED-SATELLITE (space-to-Earth). MOBILE-SATELLITE (space-to-Earth). G117

Non-Gov: FIXED-SATELLITE (space-to-Earth). MOBILE-SATELLITE (space-to-Earth).

Government satellite services in this band are limited to military systems.

## 40.5-42.5 GHz

Gov: US211

Non-Gov: BROADCASTING-SATELLITE. /BROADCASTING/. Fixed. Mobile. US211

This band is allocated primarily to, but not used for, direct broadcasting from satellites and terrestrial broadcasting.

Challenges to radio wave propagation in this part of the spectrum include the effects of rain, attenuation by water vapor and oxygen, the effects of clouds and clear air, and tropospheric scintillation.

Under experimental FCC license, tests are underway in numerous U.S. cities of TV, multimedia computer, and video program information distribution in this band.

The FCC has begun a sweeping examination of the millimeter wave bands above 40 GHz, the so-called "nosebleed" portion of the spectrum. The Commission proposed new, separate allocations for each of three categories: a Licensed Millimeter Wave Service (LMWS); low-power unlicensed devices under Part 15 rules; and vehicle radar (see 45.5–47 GHz).

"Millimeter wave technology applications could include transmission of high resolution video images, access to large databases, and communication system backbones," the FCC said. "While spectrum to accommodate wide bandwidth applications is becoming scarce below 40 GHz, the millimeter wave region of the spectrum is largely unused and can accommodate those bandwidths."

According to Texas Instruments, "Wireless networks implemented at 28 GHz, 40 GHz and higher frequencies will allow new services and applications to be brought to the consumer more rapidly at lower cost in the future. These applications, fostered and supported by new spectrum allocations, will provide two-way local area networks that make use of advanced modulation and a new video digital compression technology for spectrum-efficient digital video and other services.

"As video sources and digital databases grow, consumers will demand faster communication and data transfer media that provide the flexibility offered by wireless networks. This demand with a large consumer base will require very wide blocks of spectrum," the company said. "The development work being done in the research laboratories at 44 GHz in support of the MilSatCom Milstar program and other research being conducted at 75 and 94 GHz will lead to the next generation of components to support civil commercial spectrum applications for the digital society of tomorrow."

LMWS operations would be allocated 40.5–42.5, 47.4–48.2, 71–71.5, 84–84.5, 103–103.5, 116–116.5, 122–122.5,

126–126.5 and 152–152.5 GHz. Each LMWS band would be divided into two contiguous blocks.

The FCC believes uses of millimeter wave spectrum are likely to be similar to the Local Multipoint Distribution Service (LMDS, see 27.5–29.5 GHz). LMDS is a broadband wireless service with multiple transmitter sites. It delivers entertainment programming to consumers, and it also is capable of interactive data and voice uses.

Unlicensed devices were granted the band 59–64 GHz, and at presstime were proposed to receive 71.5–72, 84.5–85, 103.5–104, 116.5–117, 122.5–123, 126.5–127 and 152.5–153 GHz.

The FCC said its experience with Part 15 spread spectrum rules (see 902–928 and 2400–2402 MHz) and with unlicensed 2 GHz Personal Communications Services (see 2390–2400 MHz) shows that there is a potentially significant demand for unlicensed devices, some of which apply military technology to commercial uses.

The unlicensed device rules are expected to allow virtually any short-range use, with technical restrictions to minimize interference and avoid potential safety problems.

On the other hand, "given the long lead times needed to develop equipment to operate in this region of the spectrum, it does not offer a practical, near-term alternative for most unlicensed technologies," according to the Part 15 Coalition. "Moreover, due to oxygen absorption effects at higher frequencies, these bands are not well suited for medium-range, high-speed, outdoor data links, which promise to be a vital component of the National Information Infrastructure."

It seems likely that at least some millimeter wave products will have to be professionally installed, for example on towers or buildings, in order to ensure that people do not approach the energized antenna.

## 42.5-43.5 GHz

Gov: FIXED. FIXED-SATELLITE (Earth-to-space). MOBILE except aeronautical mobile. RADIO ASTRONOMY. 900

Non-Gov: FIXED. FIXED-SATELLITE (Earth-to-space). MOBILE except aeronautical mobile. RADIO ASTRONOMY. 900

This band is used for observations of silicon monoxide molecules in stars.

Active uses such as satellite uplinks and mobile radio have to protect radio astronomy from interference.

## 43.5-45.5 GHz

Gov: FIXED-SATELLITE (Earth-to-space). MOBILE-SATELLITE (Earth-to-space). G117

Non-Gov: None

The Milstar uplink is in this band (see 20.2–21.2 GHz). Many bands above 30 GHz are allocated to the Fixed Satellite Service, but only this one is presently used.

## 45.5-47 GHz

Gov: MOBILE. MOBILE-SATELLITE (Earth-to-space). RADIONAVIGATION. RADIONAVIGATION-SATELLITE. 903

Non-Gov: MOBILE. MOBILE-SATELLITE (Earth-to-space). RADIONAVIGATION. RADIONAVIGATION-SATELLITE. 903

FCC: RADIO FREQUENCY DEVICES (15)

The FCC designated 46.7–46.9 GHz for vehicle radar. Vehicle radar now is also permitted in 76–77 GHz and at presstime was to receive 94.7–95.7 and 139–140 GHz.

Vehicle radar includes forward-looking applications such as radar cruise control, which adjusts vehicle speed based on traffic ahead; collision warning; and collision avoidance which affects vehicle control and braking in response to an impending collision. Side- and rear-looking radars detect obstacles by monitoring objects less visible to the driver, and alerting the driver with visual displays or sounds.

FCC Part 15 rules permit the use of 46.7–46.9 and 76–77 GHz for data communications as long as the primary mode of operation is vehicle radar. Operation under the Part 15 vehicle radar rules in these bands is not allowed on aircraft or satellites.

## 47-47.2 GHz

Gov: None

Non-Gov: AMATEUR. AMATEUR-SATELLITE.

FCC: AMATEUR (97).

## 47.2-50.2 GHz

Gov: FIXED. FIXED-SATELLITE (Earth-to-space). MOBILE. 904 US264 US297

Non-Gov: FIXED. FIXED-SATELLITE (Earth-to-space). MOBILE. 904 US264 US297

There is no known active use of this band. Radio astronomers monitor 48.94–49.04 GHz to study molecular material in galaxies.

The FCC proposed to allocate 47.2–47.4 GHz to vehicle radars and 47.4–48.2 GHz to a new Licensed Millimeter Wave Service (see 40.5–42.5 GHz).

The government designated 49.7–50.2 GHz as a candidate segment for Intelligent Transportation Systems. It would support vehicle radar and links used for automated vehicle control ("platooning") and for coordination of evasive action between vehicles about to collide.

Sky Station International proposed an astonishing Global Stratospheric Telecommunications Service (GSTS) in the

47.2–47.5 GHz (subscriber uplink, feeder links, and platform control) and 47.9–48.2 GHz (downlink) segments.

The $4.2 billion Sky Station GSTS system would consist of 250 geostationary, airborne platforms supported by balloons. Each of the 37-ton platforms would operate from about 100,000 feet altitude, powered by solar energy and propelled by ion emission.

There are few position-disturbing forces in the stratosphere, a region above 99% of the oxygen-containing atmosphere.

GSTS would compete with broadband satellite offerings via a network connecting balloon platforms, ground facilities and subscriber communicators. Each platform can cover 400,000 square miles and would be capable of creating 2100 separate geographical cells.

Platforms will use "multiple redundant safety features that will eliminate the risk of injury or harm to airborne vehicles and Earth inhabitants," Sky Station said. "These safety features includes multiply-redundant balloon modules (only half of which are sufficient to ensure a gentle and controlled landing), back-up parachute systems, multiple motion sensors (radar altimeters, accelerometers and GPS devices) and compatibility with helicopter recovery operations. ...Damage on Earth is no more likely to occur than from satellite launch and de-orbit operations."

The company plans to begin launching platforms at the rate of one per month in 1999. Its business plan calls for 18 million subscribers after five years of service, each paying ten cents a minute for telephone, videophone and data services including wireless Internet connection at 64 kbps. Sky Station believes that the GSTS could accommodate as many as ten different FCC licensees.

Motorola, proponent of the Iridium project, argued to the FCC that the requested spectrum would be needed for expansion of the Fixed Satellite Service and that GSTS could threaten FSS operations.

Foreign countries would likely view GSTS as a "blatant attempt at infringement of their sovereign rights" to airspace, Motorola said. Sky Station replied that "the stratosphere is a region of proto-space that the United States and other countries have steadfastly refused to define as part of any nation's sovereign airspace."

Hughes Communications took a similar position. "Although the 47.2–47.5 GHz and 47.9–48.2 GHz bands are not commercially used today for the Fixed Satellite Service or Broadcasting Satellite Service, these bands provide much needed expansion capacity to accommodate future use requirements of those services.

"The 40 GHz band provides a basis for significant economic growth and new jobs in the satellite sector for spacecraft manufacturers, launch providers, service providers, and ground equipment manufacturers alike, but that expansion can occur only if sufficient spectrum remains available for the FSS and BSS."

## 50.2-50.4 GHz

Gov: EARTH EXPLORATION-SATELLITE (passive). FIXED. MOBILE. SPACE RESEARCH (passive). US263

Non-Gov: EARTH EXPLORATION-SATELLITE (passive). FIXED. MOBILE. SPACE RESEARCH (passive). US263

Ostensibly this area of the spectrum is used for satellite atmospheric temperature sensing. NASA found, however, that allocations in 50.2–66 GHz are currently not optimum and cannot provide temperature data in the mesosphere. A band between 60.3 and 61.3 GHz has been identified as necessary for mesospheric temperature measurements.

## 50.4-51.4 GHz

Gov: FIXED. FIXED-SATELLITE (Earth-to-space). MOBILE. MOBILE-SATELLITE (Earth-to-space). G117

Non-Gov: FIXED. FIXED-SATELLITE (Earth-to-space). MOBILE. MOBILE-SATELLITE (Earth-to-space).

U.S. satellites in this band are limited to military systems. This band was designated a candidate for Intelligent Transportation Systems (see 47.2–50.2 GHz).

## 51.4-54.25 GHz

Gov: EARTH EXPLORATION-SATELLITE (passive). RADIO ASTRONOMY. SPACE RESEARCH (passive). US246

Non-Gov: EARTH EXPLORATION-SATELLITE (passive). RADIO ASTRONOMY. SPACE RESEARCH (passive). US246

No transmissions are permitted in this receive-only scientific band (see 50.2–50.4 GHz).

## 54.25-58.2 GHz

Gov: EARTH EXPLORATION-SATELLITE (passive). FIXED. INTER-SATELLITE. MOBILE. SPACE RESEARCH (passive). 909 US263

Non-Gov: EARTH EXPLORATION-SATELLITE (passive). FIXED. INTER-SATELLITE. MOBILE. SPACE RESEARCH (passive). 909 US263

In addition to its use for satellite sensors that measure temperatures in layers of the atmosphere, this band is one of those proposed for intersatellite links in broadband fixed satellites (see 17.7–17.8 GHz).

NASA is concerned that intersatellite use of this band would interfere with spaceborne sensors. "The result would be an electromagnetic screen that would prevent acquisition of vital environmental data," the agency said.

## 58.2-59 GHz

Gov: EARTH EXPLORATION-SATELLITE (passive). RADIO ASTRONOMY. SPACE RESEARCH (passive). US246

Non-Gov: EARTH EXPLORATION-SATELLITE (passive). RADIO ASTRONOMY. SPACE RESEARCH (passive). US246

No transmissions are permitted in this receive-only scientific band.

## 59-64 GHz

Gov: FIXED. INTER-SATELLITE. MOBILE. RADIOLOCATION. 909 910 911

Non-Gov: FIXED. INTER-SATELLITE. MOBILE. RADIOLOCATION. 909 910 911

FCC: Radio frequency devices (15)

The propagation range in this band is extremely limited due to effects on radio waves by oxygen in the atmosphere. This attenuation is significant throughout the millimeter wave spectrum, but increases around 60 and 120 GHz.

This limited range of transmission permits a high concentration of transmitters to be used in an area. The FCC designated this band for unlicensed "novel broadband applications such as wireless computer-to-computer communications" (see 40.5–42.5 GHz).

Not all uses will be for communications. Industrial safety and control sensors will use this band. It also is one of those proposed for use by intersatellite links in broadband fixed satellites (see 17.7–17.8 GHz).

The National Oceanic and Atmospheric Administration and the Air Force plan to use 60.4–61.2 GHz for weather satellite passive sensing of temperature.

The 61–61.5 GHz segment is available to Industrial, Scientific and Medical devices.

## 64-65 GHz

Gov: EARTH EXPLORATION-SATELLITE (passive). RADIO ASTRONOMY. SPACE RESEARCH (passive). US246

Non-Gov: EARTH EXPLORATION-SATELLITE (passive). RADIO ASTRONOMY. SPACE RESEARCH (passive). US246

No transmissions are permitted in this receive-only scientific band.

## 65-66 GHz

Gov: EARTH EXPLORATION-SATELLITE. SPACE RESEARCH. Fixed. Mobile.

Non-Gov: EARTH EXPLORATION-SATELLITE. SPACE RESEARCH. Fixed. Mobile.

This is another satellite atmospheric temperature sensor band.

## 66-71 GHz

Gov: MOBILE. MOBILE-SATELLITE. RADIONAVIGATION. RADIONAVIGATION-SATELLITE. 903

Non-Gov: MOBILE. MOBILE-SATELLITE. RADIONAVIGATION. RADIONAVIGATION-SATELLITE. 903

The footnotes provide for feeder link use of this band in the mobile satellite services. Aviation radar in this band includes an inclement weather guidance system. The 68.9–72.7 GHz spectrum is a candidate for Intelligent Transportation Systems (see 47.2–50.2 GHz).

## 71-74 GHz

Gov: FIXED. FIXED-SATELLITE (Earth-to-space). MOBILE. MOBILE-SATELLITE (Earth-to-space). US270

Non-Gov: FIXED. FIXED-SATELLITE (Earth-to-space). MOBILE. MOBILE-SATELLITE (Earth-to-space). US270

The FCC proposed to allocate 71–71.5 GHz to a Licensed Millimeter Wave Service and 71.5–72 GHz to low-power unlicensed devices (see 40.5–42.5 GHz).

## 74-75.5 GHz

Gov: FIXED. FIXED-SATELLITE (Earth-to-space). MOBILE. US297

Non-Gov: FIXED. FIXED-SATELLITE (Earth-to-space). MOBILE. US297

## 75.5-76 GHz

Gov: None

Non-Gov: AMATEUR. AMATEUR-SATELLITE.

FCC: AMATEUR (97).

## 76-77 GHz

Gov: RADIOLOCATION.

Non-Gov: RADIOLOCATION. Amateur.

FCC: RADIO FREQUENCY DEVICES (15)

This is one of the vehicle radar bands (see 45.5–47 GHz). The FCC proposed to suspend Amateur Radio use of this band until it determines that amateur transmissions will not interfere with vehicle radar and pose a safety threat.

FCC Part 15 rules permit the use of 46.7–46.9 and 76–77 GHz for data communications as long as the primary mode of operation is vehicle radar.

## 77-81 GHz

Gov: 912

Non-Gov: RADIOLOCATION. Amateur. Amateur-Satellite. 912

FCC: Amateur (97).

The 1992 World Administrative Radio Conference allocated 76–81 GHz on a secondary basis to the Space Research Service, intended for cloud radars in 78–79 GHz. The FCC proposed to adopt this allocation domestically. As it turns out, however, cloud radars—which collect data for climate study—work better at 95 GHz than at 79 GHz. Cloud particles reflect twice as much energy back to the radar at the higher frequency.

The Defense Department is developing exotic systems in 80–120 GHz, including quasi-optical gyrotrons for material processing and traveling wave tubes for countermeasures.

## *81-84 GHz*

Gov: FIXED. FIXED-SATELLITE (space-to-Earth). MOBILE. MOBILE-SATELLITE (space-to-Earth).

Non-Gov: FIXED. FIXED-SATELLITE (space-to-Earth). MOBILE. MOBILE-SATELLITE (space-to-Earth).

## *84-86 GHz*

Gov: FIXED. MOBILE. 913 US211

Non-Gov: FIXED. MOBILE. BROADCASTING. BROADCASTING-SATELLITE. 913 US211

The FCC proposed to allocate 84–84.5 GHz to a Licensed Millimeter Wave Service and 84.5–85 GHz to low-power unlicensed devices (see 40.5–42.5 GHz).

## *86-92 GHz*

Gov: EARTH EXPLORATION-SATELLITE (passive). RADIO ASTRONOMY. SPACE RESEARCH (passive). US74 US246

Non-Gov: EARTH EXPLORATION-SATELLITE (passive). RADIO ASTRONOMY. SPACE RESEARCH (passive). US74 US246

No transmissions are permitted in this widely used radio astronomy band.

Because there is little energy absorption by atmospheric water and oxygen, this band is excellent for observations of celestial objects. It is also used by satellite instruments that observe water vapor, clouds, ice and snow.

## 92-95 GHz

Gov: FIXED. FIXED-SATELLITE (Earth-to-space). MOBILE. RADIOLOCATION. 914

Non-Gov: FIXED. FIXED-SATELLITE (Earth-to-space). MOBILE. RADIOLOCATION. 914

The FCC proposed to allocate 94.7–95.7 GHz to vehicle collision-avoidance radars (see 40.5–42.5 GHz). It is used in light rail radar that detects people or objects in the train path and alerts the operator.

Aviation radar in this band includes inclement weather guidance and imaging of the runway, even if obscured by fog or smoke. This band is much researched by defense users, especially for active radar in missiles.

Millimeter wave imaging can display hidden contraband or weapons, including nonmetal objects, for such uses as airport security. It detects thermal emissions and reflections from materials. The object appears in the 94 GHz image as a shadow against the body. Operators may have to be the same sex as the person being screened, because the technology sees through clothing.

## 95-100 GHz

Gov: MOBILE. MOBILE-SATELLITE. RADIONAVIGATION. RADIONAVIGATION-SATELLITE. Radiolocation. 902 903 904

Non-Gov: MOBILE. MOBILE-SATELLITE. RADIONAVIGATION. RADIONAVIGATION-SATELLITE. Radiolocation. 902 903 904

The FCC proposed to allocate 94.7–95.7 GHz to vehicle collision-avoidance radars (see 40.5–42.5 GHz).

## 100-102 GHz

Gov: EARTH EXPLORATION-SATELLITE (passive). SPACE RESEARCH (passive). 722 US246

Non-Gov: EARTH EXPLORATION-SATELLITE (passive). SPACE RESEARCH (passive). 722 US246

No transmissions are permitted in this receive-only scientific band, used in satellite remote sensing and radio astronomy.

## 102-105 GHz

Gov: FIXED. FIXED-SATELLITE (space-to-Earth). 722 US211

Non-Gov: FIXED. FIXED-SATELLITE (space-to-Earth). 722 US211

The FCC proposed to allocate 103–103.5 GHz to a Licensed Millimeter Wave Service and 103.5–104 GHz to unlicensed use (see 40.5–42.5 GHz).

## 105-116 GHz

Gov: EARTH EXPLORATION-SATELLITE (passive). RADIO ASTRONOMY. SPACE RESEARCH (passive). 722 US74 US246

Non-Gov: EARTH EXPLORATION-SATELLITE (passive). RADIO ASTRONOMY. SPACE RESEARCH (passive). 722 US74 US246

No transmissions are permitted in this receive-only scientific band, used in satellite remote sensing and radio astronomy.

## 116-126 GHz

Gov: EARTH EXPLORATION-SATELLITE (passive). FIXED. INTER-SATELLITE. MOBILE. SPACE RESEARCH (passive). 722 909 915 916 US211 US263

Non-Gov: EARTH EXPLORATION-SATELLITE (passive). FIXED. INTER-SATELLITE. MOBILE. SPACE RESEARCH (passive). 722 909 915 916 US211 US263

The FCC proposed to allocate 116–116.5 and 122--122.5 GHz to a Licensed Millimeter Wave Service and 116.5–117 and 122.5–123 GHz to unlicensed use (see 40.5–42.5 GHz).

Remote sensing satellites use this band to measure atmospheric temperature. Astronomers use it to measure ozone, carbon monoxide and nitrous oxide. Scientists are concerned about FCC proposals to allocate the band to new uses because of interference to remote sensing.

Amateur Radio is allocated 119.98–120.02 GHz on a secondary basis. The 122–123 GHz segment is designated for Industrial, Scientific and Medical applications.

## 126-134 GHz

Gov: FIXED. INTER-SATELLITE. MOBILE. RADIOLOCATION 909 910

Non-Gov: FIXED. INTER-SATELLITE. MOBILE. RADIOLOCATION 909 910

The FCC proposed to allocate 126–126.5 GHz to a Licensed Millimeter Wave Service and 126.5–127 GHz to unlicensed use (see 40.5–42.5 GHz).

## 134-142 GHz

Gov: MOBILE. MOBILE-SATELLITE. RADIONAVIGATION. RADIONAVIGATION-SATELLITE. Radiolocation. 902 903 917 918

Non-Gov: MOBILE. MOBILE-SATELLITE. RADIONAVIGATION. RADIONAVIGATION-SATELLITE. Radiolocation. 902 903 917 918

The FCC proposed to allocate 139–140 GHz to vehicle anti-collision radars (see 40.5–42.5 GHz). Military tracking and navigation radars operate around 140 GHz.

## 142-144 GHz

Gov: None

Non-Gov: AMATEUR. AMATEUR-SATELLITE.

FCC: Amateur (97).

## 144-149 GHz

Gov: RADIOLOCATION. 918

Non-Gov: RADIOLOCATION. Amateur. Amateur-Satellite. 918

FCC: Amateur (97).

The 148.5–151.5 GHz spectrum and nearby bands are used for satellite remote sensing of atmospheric moisture.

## 149-150 GHz

Gov: FIXED. FIXED-SATELLITE (space-to-Earth). MOBILE.

Non-Gov: FIXED. FIXED-SATELLITE (space-to-Earth). MOBILE.

## 150-151 GHz

Gov: EARTH EXPLORATION-SATELLITE (passive). FIXED. FIXED-SATELLITE (space-to-Earth). MOBILE. SPACE RESEARCH (passive). 919 US263

Non-Gov: EARTH EXPLORATION-SATELLITE (passive). FIXED. FIXED-SATELLITE (space-to-Earth). MOBILE. SPACE RESEARCH (passive). 919 US263

## 151-164 GHz

Gov: FIXED. FIXED-SATELLITE. US211

Non-Gov: FIXED. FIXED-SATELLITE. US211

The FCC proposed to allocate 152–152.5 GHz to a Licensed Millimeter Wave Service and 152.5–153 GHz to unlicensed use (see 40.5–42.5 GHz).

The Big Bang produced cosmic background radiation that peaks around 160 GHz (see 217–231 GHz).

## *164-168 GHz*

Gov: EARTH EXPLORATION-SATELLITE (passive). RADIO ASTRONOMY. SPACE RESEARCH (passive). US246

Non-Gov: EARTH EXPLORATION-SATELLITE (passive). RADIO ASTRONOMY. SPACE RESEARCH (passive). US246

No transmissions are permitted in this receive-only scientific band. Satellites use it to detect terrestrial water vapor and chlorine oxide.

## *168-170 GHz*

Gov: FIXED. MOBILE.

Non-Gov: FIXED. MOBILE.

The federal government designated 168–168.3 GHz as a candidate segment for Intelligent Transportation Systems (see 47.2–50.2 GHz).

## *170-174.5 GHz*

Gov: FIXED. INTER-SATELLITE. MOBILE. 909 919

Non-Gov: FIXED. INTER-SATELLITE. MOBILE. 909 919

## *174.5-176.5 GHz*

Gov: EARTH EXPLORATION-SATELLITE (passive). FIXED. INTER-SATELLITE. MOBILE. SPACE RESEARCH (passive). 909 919 US263

Non-Gov: EARTH EXPLORATION-SATELLITE (passive). FIXED. INTER-SATELLITE. MOBILE. SPACE RESEARCH (passive). 909 919 US263

Remote sensing satellites detect nitrous oxide in this band.

## 176.5-182 GHz

Gov: FIXED. INTER-SATELLITE. MOBILE. 909 919 US211

Non-Gov: FIXED. INTER-SATELLITE. MOBILE. 909 919 US211

## 182-185 GHz

Gov: EARTH EXPLORATION-SATELLITE (passive). RADIO ASTRONOMY. SPACE RESEARCH (passive). US246

Non-Gov: EARTH EXPLORATION-SATELLITE (passive). RADIO ASTRONOMY. SPACE RESEARCH (passive). US246

No transmissions are permitted in this receive-only scientific band. Satellite instruments use it to measure atmospheric water vapor, clouds, precipitation and ozone.

## 185-190 GHz

Gov: FIXED. INTER-SATELLITE. MOBILE. 909 919 US211

Non-Gov: FIXED. INTER-SATELLITE. MOBILE. 909 919 US211

## 190-200 GHz

Gov: MOBILE. MOBILE-SATELLITE. RADIONAVIGATION. RADIONAVIGATION-SATELLITE. 722 902 903

Non-Gov: MOBILE. MOBILE-SATELLITE. RADIONAVIGATION. RADIONAVIGATION-SATELLITE. 722 902 903

## 200-202 GHz

Gov: EARTH EXPLORATION-SATELLITE (passive). FIXED. MOBILE. SPACE RESEARCH (passive). 722 US263

Non-Gov: EARTH EXPLORATION-SATELLITE (passive). FIXED. MOBILE. SPACE RESEARCH (passive). 722 US263

Remote sensing satellites detect nitrous oxide in this band.

## 202-217 GHz

Gov: FIXED. FIXED-SATELLITE (Earth-to-space). MOBILE. 722

Non-Gov: FIXED. FIXED-SATELLITE (Earth-to-space). MOBILE. 722

## 217-231 GHz

Gov: EARTH EXPLORATION-SATELLITE (passive). RADIO ASTRONOMY. SPACE RESEARCH (passive). 722 US74 US246

Non-Gov: EARTH EXPLORATION-SATELLITE (passive). RADIO ASTRONOMY. SPACE RESEARCH (passive). 722 US74 US246

No transmissions are permitted in this receive-only scientific band, one of the most important to millimeter wave radio astronomy.

This band is used to detect carbon monoxide, nitrous oxide, and other complex molecules in gas within galaxies.

It also is used in studies of cosmic background radiation, the broadband radio noise remaining from the alleged origin of the universe (the Big Bang). The peak of the radiation is near 160 GHz. Astronomers use this nearby 217–231 GHz allocation to observe it. Such observations are used to test theories about the origin of the cosmos.

For Earth remote sensing, a 220 GHz receiver, part of a millimeter wave imaging radiometer, measures atmospheric water vapor, clouds and precipitation in the Defense Meteorological Satellite Program (see 1690–1700 MHz).

## 231-235 GHz

Gov: FIXED. FIXED-SATELLITE (space-to-Earth). MOBILE. Radiolocation. US211

Non-Gov: FIXED. FIXED-SATELLITE (space-to-Earth). MOBILE. Radiolocation. US211

## 235-238 GHz

Gov: EARTH EXPLORATION-SATELLITE (passive). FIXED. FIXED-SATELLITE (space-to-Earth). MOBILE. SPACE RESEARCH (passive). US263

Non-Gov: EARTH EXPLORATION-SATELLITE (passive). FIXED. FIXED-SATELLITE (space-to-Earth). MOBILE. SPACE RESEARCH (passive). US263

Remote sensing satellites detect ozone in this band.

## 238-241 GHz

Gov: FIXED. FIXED-SATELLITE (space-to-Earth). MOBILE. Radiolocation.

Non-Gov: FIXED. FIXED-SATELLITE (space-to-Earth). MOBILE. Radiolocation.

## 241-248 GHz

Gov: RADIOLOCATION. 922

Non-Gov: RADIOLOCATION. Amateur. Amateur-Satellite. 922

FCC: Amateur (97).

Industrial, Scientific and Medical devices may use the 244–246 GHz segment.

## 248-250 GHz

Gov: None

Non-Gov: AMATEUR. AMATEUR-SATELLITE.

FCC: AMATEUR (97).

## 250-252 GHz

Gov: EARTH EXPLORATION-SATELLITE (passive). SPACE RESEARCH (passive). 923

Non-Gov: EARTH EXPLORATION-SATELLITE (passive). SPACE RESEARCH (passive). 923

Remote sensing satellites detect nitrous oxide in this band.

## 252-265 GHz

Gov: MOBILE. MOBILE-SATELLITE. RADIONAVIGATION. RADIONAVIGATION-SATELLITE. 902 903 923 924 US211

Non-Gov: MOBILE. MOBILE-SATELLITE. RADIONAVIGATION. RADIONAVIGATION-SATELLITE. 902 903 923 924 US211

## 265-275 GHz

Gov: FIXED. FIXED–SATELLITE (Earth-to-space). MOBILE. RADIO ASTRONOMY. 926

Non-Gov: FIXED. FIXED-SATELLITE (Earth-to-space). MOBILE. RADIO ASTRONOMY. 926

## 275-300 GHz

Gov: FIXED. MOBILE. 927

Non-Gov: FIXED. MOBILE. 927

Frequencies above 300 GHz are not yet allocated by the United States. However, research is taking place above 300 GHz into radio technology, especially for applications such as remote sensing, radiolocation and radio astronomy. NASA's Submillimeter Wave Astronomy Satellite will examine astrochemical phenomena around 500 GHz.

International Footnote 927 permits experiments in 275–400 GHz for "various active and passive services." It recommends that special care be taken to protect from interference the passive services such as radio astronomy, the Space Research Service (passive) and the Earth exploration-satellite service (passive).

# INTERNATIONAL FOOTNOTES

The following International Footnotes are those cited in the U.S. Table of Allocations for bands between 30 MHz and 300 GHz. These are a subset of all International Footnotes drawn from the Radio Regulations of the International Telecommunication Union.

501 The carrier frequencies 2182 kHz, 3023 kHz, 5680 kHz, 8364 kHz, and the frequencies 121.5 MHz, 156.8 MHz and 243 MHz may also be used, in accordance with the procedures in force for terrestrial radiocommunication services, for search and rescue operations concerning manned space vehicles. The conditions for the use of the frequencies are prescribed in Articles 38 and N 38. The same applies to the frequencies 10003 kHz, 14993 kHz and 19993 kHz, but in each of these cases emissions must be confined in a band of ± 3 kHz about the frequency.

510 For the use of these bands allocated to the amateur service at 3.5 MHz, 7.0 MHz, 10.1 MHz, 14.0 MHz, 18.068 MHz, 21.0 MHz, 24.89 MHz and 144 MHz in the event of natural disasters, see Resolution 640.

547 In making assignments to stations of other services to which the band 37.5–38.25 MHz is allocated, administrations are urged to take all practicable steps to protect the radio astronomy service from harmful interference. Emissions from space or airborne stations can be particularly serious sources of interference to the radio astronomy service (see Nos. 343 and 344 and Article 36).

548 The band 40.66–40.70 MHz (centre frequency 40.68 MHz) is designated for industrial, scientific and medical (ISM) applications. Radiocommunication services operating within this band must accept harmful interference which may be caused by these applications. ISM equipment operating in this band is subject to the provisions of No. 1815.

572 Additional allocation: The frequency 75 MHz is assigned to aeronautical marker beacons. Administrations shall refrain from assigning frequencies close to the limits of the guardband to stations of other services which, because of their power or geographical position, might cause harmful interference or otherwise place a constraint on marker beacons. Until 31 December 1989, administrations in Regions 2 and 3 should refrain from assigning frequencies to other services in the bands 74.6–74.8 MHz and 75.2–75.4 MHz. In the future every effort should be made to improve further the characteristics of airborne receivers and to limit the power of transmitting stations close to the limits 74.8 and 75.2 MHz.

576 Different category of service: in the United States, the French Overseas Departments in Region 2, Guyana, Jamaica, Mexico and Paraguay, the allocation of the band 76-88 MHz to the fixed and mobile services is on a primary basis (see No. 425).

591 Subject to agreement obtained under the procedure set forth in article 14, the band 117.975–137 MHz is also allocated to the aeronautical mobile-satellite (R) service on a secondary basis and on the condition that harmful interference is not caused to the aeronautical mobile (R) service.

592 The bands 121.45–121.55 MHz and 242.95–243.05 MHz are also allocated to the mobile-satellite service for the reception on board satellites of emissions from emergency position-indicating radiobeacons transmitting at 121.5 MHz and 243 MHz.

593 In the band 117.975–136 MHz, the frequency 121.5 MHz is the aeronautical emergency frequency and, where required, the frequency 123.1 MHz is the aeronautical frequency auxiliary to 121.5 MHz. Mobile stations of the maritime mobile service may communicate on these frequencies under the conditions laid down in Articles 38

and N 38 for distress and safety purposes with stations of the aeronautical mobile service.

599A The use of the band 137–138 MHz by the mobile-satellite service is subject to the application of the coordination and notification procedures set forth in RES46 (WARC-92). However, coordination of a space station of the mobile-satellite service with respect to terrestrial services is required only if the power flux-density produced by the station exceeds -125 dB(W/m$^2$/4 kHz) at the Earth's surface. The above power flux-density limit shall apply until such time as a competent world administrative radio conference revises it. In making assignments to the space stations in the mobile-satellite service in the above band, administrations shall take all practicable steps to protect the radio astronomy service in the 150.05–153 MHz band from harmful interference from unwanted emissions.

599B The use of the bands 137–138 MHz, 148–149.9 MHz and 400.15–401 MHz by the mobile-satellite service and the band 149.9–150.05 MHz by the land mobile-satellite service is limited to non-geostationary-satellite systems.

608 Subject to agreement obtained under the procedure set forth in Article 14, the band 148–149.9 MHz may be used by the space operation service (Earth-to-space). The bandwidth of an individual transmission shall not exceed ± 25 kHz.

608A The use of the band 148–149.9 MHz by the mobile-satellite service is subject to the application of the coordination and notification procedures set forth in Resolution 46. The mobile-satellite service shall not constrain the development and use of fixed, mobile and space operation services in the band 148–149.9 MHz. Mobile earth stations in the mobile-satellite service shall not produce a power flux-density in excess of -150 dB(W/m$^2$/4 kHz) outside national boundaries.

608B The use of the band 149.9–150.05 MHz by the land mobile-satellite service is subject to the application of the coordination and notification procedures set forth in Resolution 46. The land mobile-satellite service shall not constrain the development and use of the radionavigation-satellite service in the band 149.9–150.05 MHz. Land mobile earth stations of the mobile-satellite service shall not produce power flux-density in excess of -150 dB(W/m$^2$/4 kHz) outside national boundaries.

609A Recognizing that the use of the band 149.9–150.05 MHz by the fixed and mobile services may cause harmful interference to the radionavigation-satellite service, administrations are urged not to authorize such use in application No. 342.

613 The frequency 156.8 MHz is the international distress, safety and calling frequency for the maritime mobile VHF radiotelephone service. The conditions for the use of this frequency are contained in Articles 38 and N 38. In the bands 156–156.7625 MHz, 156.8375–157.45 MHz, 160.6–160.975 MHz and 161.475–162.05 MHz, each administration shall give priority to the maritime mobile service on only such frequencies as are assigned to stations of the maritime mobile service by the administration (see Articles 38, N 38 and 60). Any use of frequencies in these bands by station of other services to which they are allocated should be avoided in areas where such use might cause harmful interference to the maritime mobile VHF radiocommunication service. However, the frequency 156.8 MHz and the frequency bands in which priority is given to the maritime mobile service may be used for radiocommunications on inland waterways subject to agreement between interested and affected administrations and taking into account current frequency usage and existing agreements.

613A In the maritime mobile VHF service the frequency 156.525 MHz is to be used exclusively for digital selective calling for distress, safety and calling (see Resolution 323 (Mob-87)). The conditions for the use of this frequency are prescribed in Articles 38, N 38 and 60 and in Appendix 18.

627 In Region 2, no new stations in the radiolocation service may be authorized in the band 216–225 MHz. Stations authorized prior to 1 January 1990 may continue to operate on a secondary basis.

642 The frequency 243 MHz is the frequency in this band for use by survival craft stations and equipment used for survival purposes (see Article 38).

644 In making assignments to stations of other services to which the band 322–328.6 MHz is allocated, administrations are urged to take all practicable steps to protect the radio astronomy service from harmful interference. Emissions from space or airborne stations can be particularly serious sources of interference to the radio astronomy service (see Nos. 343 and 344 and Article 36).

645 Limited to Instrument Landing Systems (glide path).

645B Recognizing that the use of the band 399.9–400.05 MHz by the fixed and mobile service may cause harmful interference to the radionavigation satellite service, administrations are urged not to authorize such use in application of No. 342.

646 Emissions shall be confined in a band of ± 25 kHz about the standard frequency 400.1 MHz.

647 Additional allocation: In Afghanistan, Saudi Arabia, Austria, Bahrain, Bulgaria, Colombia, Costa Rica, Cuba, Egypt, the United Arab Emirates, Ecuador, Hungary, Indonesia, Iran, Iraq, Israel, Kuwait, Liberia, Malaysia, Nigeria, Oman, Pakistan, the Philippines, Poland, Qatar, Syria, the German Democratic Republic, Roumania, Singapore, Somalia, Sri Lanka, Czechoslovakia, Thailand, the U.S.S.R. and Yugoslavia, the band 400.05–401 MHz is also allocated to the fixed and mobile services on a primary

basis.

647A The band 400.15–401 MHz is also allocated to the space research service in the space-to-space direction for communications with manned space vehicles. In this application, the space research service will not be regarded as a safety service.

647B The use of the band 400.15–401 MHz by the mobile-satellite service is subject to the application of the coordination and notification procedures set forth in Resolution 46. However, coordination of a space station of the mobile-satellite service with respect to terrestrial services is required only if the power flux-density produced by the station exceeds -125 dB(W/$m^2$/4 kHz) at the Earth's surface. The above power flux-density limit shall apply until such time as a competent world administrative radio conference revises it. In making assignments to the space stations in the mobile-satellite service in the above band, administrations shall take all practicable steps to protect the radio astronomy service in the band 406.1–410 MHz from harmful interference from unwanted emissions.

649 The use of the band 406–406.1 MHz by the mobile-satellite service is limited to low power satellite emergency position-indicating radiobeacons (see also Articles 38 and N 38).

649A Any emission capable of causing harmful interference to the authorized uses of the band 406–406.1 MHz is prohibited.

651A Use of the band 410–420 MHz by the space research service is limited to communications within 5 km of an orbiting, manned space vehicle.

664 In the bands 435–438 MHz, 1260–1270 MHz, 2400–2450 MHz, 3400–3410 MHz (in Regions 2 and 3 only) and 5650–5670 MHz, the amateur-satellite service may operate subject to not causing harmful interference to other services operating in accordance with the Table (see No. 435).

Administrations authorizing such use shall ensure that any harmful interference caused by emissions from a station in the amateur-satellite service is immediately eliminated in accordance with the provisions of No. 2741. The use of the bands 1260–1270 MHz and 5650–5670 MHz by the amateur-satellite service is limited to the Earth-to-space direction.

668 Subject to agreement obtained under the procedure set forth in Article 14, the band 449.75–450.25 MHz may be used for the space operation service (Earth-to-space) and the space research service (Earth-to-space).

669 In the maritime mobile service, the frequencies 457.525 MHz, 457.550 MHz, 457.575 MHz, 467.525 MHz, 467.550 MHz and 467.575 MHz may be used by on-board communication stations. The use of these frequencies in territorial waters may be subject to the national regulations of the administration concerned. The characteristics of the equipment used shall conform to those specified in Appendix 20.

670 In the territorial waters of Canada, the United States and the Philippines, the preferred frequencies for use by on-board communication stations shall be 457.525 MHz, 457.550 MHz, 457.575 MHz and 457.600 MHz paired, respectively, with 467.750 MHz, 467.775 MHz, 467.800 MHz and 467.825 MHz. The characteristics of the equipment used shall conform to those specified in Appendix 20.

671 Earth exploration-satellite service applications, other than the meteorological-satellite service, may also be used in the bands 460–470 MHz and 1690–1710 MHz for space-to-Earth transmissions subject to not causing harmful interference to stations operating in accordance with the Table.

707 In Region 2, the band 902–928 MHz (centre frequency 915 MHz) is designated for industrial, scientific

and medical (ISM) applications. Radiocommunication service operating within this band must accept harmful interference which may be caused by these applications. ISM equipment operating in this band is subject to the provisions of No. 1815.

709 The band 960–1215 MHz is reserved on a worldwide basis for the use and development of airborne electronic aids to air navigation and any directly associated ground-based facilities.

713 In the bands 1215–1300 MHz, 3100–3300 MHz, 5250–5350 MHz, 8550–8650 MHz, 9500–9800 MHz and 13.4–14.0 GHz, radiolocation stations installed on spacecraft may also be employed for the earth exploration-satellite and space research services on a secondary basis.

714 Additional allocation: In Canada and the United States the bands 1240–1300 MHz and 1350–1370 MHz are also allocated to the aeronautical radionavigation service on a primary basis.

717 The use of the bands 1300–1350 MHz, 2700–2900 MHz and 9000–9200 MHz by the aeronautical radionavigation service is restricted to ground-based radars and to associated airborne transponders which transmit only on frequencies in these bands and only when actuated by radars operating in the same band.

718 In making assignments to stations of other services, administrations are urged to take all practicable steps to protect the spectral line observations of the radio astronomy service from harmful interference in the band 1330–1400 MHz. Emissions from space or airborne stations can be particularly serious sources of interference to the radio astronomy service (see Nos. 343 and 344 and Article 36).

720 The bands 1370–1400 MHz, 2640–2655 MHz, 4950–4990 MHz and 15.20–15.35 GHz are also allocated to the space research (passive) and earth exploration-satellite (passive) services on a secondary basis.

722 In the bands 1400–1727 MHz, 101–120 GHz and 197–220 GHz, passive research is being conducted by some countries in a programme for the search for intentional emissions of extra-terrestrial origin.

726A The bands 1525–1544 MHz, 1545–1559 MHz, 1626.5–1645.5 MHz and 1646.5–1660.5 MHz shall not be used for feeder links of any service. In exceptional circumstances, however, an earth station at a specified fixed point in any of the mobile-satellite services may be authorized by an administration to communicate via space stations using these bands.

727A The use of the band 1544–1545 MHz by the mobile-satellite service (space-to-Earth) is limited to distress and safety communications (see Article N 38).

731E The use of the band 1610–1626.5 MHz by the mobile-satellite service (Earth-to-space) and by the radiodetermination-satellite service (Earth-to-space) is subject to the application of the coordination and notification procedures set forth in Resolution 46 (WARC-92). A mobile earth station operating in either of the services in this band shall not produce an e.i.r.p. density in excess of -15 dB (W/4 kHz) in the part of the band used by systems operating in accordance with the provisions of No. 732, unless otherwise agreed by the affected administrations. In the part of the band where such systems are not operating, a value of -3 dB (W/4 kHz) is applicable. Stations of the mobile-satellite service shall not cause harmful interference to, or claim protection from, stations in the aeronautical radionavigation service, stations operating in accordance with the provisions of No. 732 and stations in the fixed service operating in accordance with the provisions of No. 730.

731F The use of the band 1613.8–1626.5 MHz by the mobile-satellite service (space-to-Earth) is subject to the application of the coordination and notification procedures set forth in Resolution 46 (WARC-92).

732 The band 1610–1626.5 MHz is reserved on a worldwide basis for the use and development of airborne electronic aids to air navigation and any directly associated ground-based or satellite-borne facilities. Such satellite use is subject to agreement obtained under the procedure set forth in Article 14.

733 The band 1610–1626.5 MHz, 5000–5250 MHz and 15.4–15.7 GHz are also allocated to the aeronautical mobile-satellite (R) service on a primary basis. Such use is subject to agreement obtained under the procedure set forth in Article 14.

733A With respect to the radiodetermination-satellite and mobile-satellite services the provisions of No. 953 do not apply in the frequency band 1610–1626.5 MHz.

733E Harmful interference shall not be caused to stations of the radio astronomy service using the band 1610.6–1613.8 MHz by stations of the radiodetermination-satellite and mobile-satellite services. (No. 2904 applies.)

734 In making assignments to stations of other services, administrations are urged to take all practicable steps to protect the radio astronomy service in the band 1610.6–1613.8 MHz from harmful interference. Emissions from space or air-borne stations can be particularly serious sources of interference to the radio astronomy service (see Nos. 343 and 344 and Article 36).

734B The use of the band 1645.5–1646.5 MHz by the mobile satellite service (Earth-to-space) and for inter-satellite links is limited to distress and safety communications (see Article N 38).

736 In making assignments to stations of other services to which the band 1660–1670 MHz is allocated, administrations are urged to take all practicable steps to protect the radio astronomy service from harmful interference. Emissions from space or airborne stations can

be particularly serious sources of interference to the radio astronomy service (see Nos. 343 and 344 and Article 36).

750A Administrations are urged to take all practicable measures to ensure that space-to-space transmissions between two or more non-geostationary satellites in the space research, space operation and Earth exploration-satellite services in the bands 2025-2110 MHz and 2200-2290 MHz, shall not impose any constraints on Earth-to-space, space-to-Earth and other space-to-space transmissions of those services and in those bands between geostationary and non-geostationary satellites.

751B Space stations of the broadcasting-satellite service in the band 2310–2360 MHz operating in accordance with No. 750B that may affect services to which this band is allocated in other countries shall be coordinated and notified in accordance with Resolution 33. Complementary terrestrial broadcasting stations shall be subject to bilateral coordination with neighboring countries prior to their bringing into use.

752 The band 2400–2500 MHz (centre frequency 2450 MHz) is designated for industrial, scientific and medical (ISM) applications. Radio services operating within this band must accept harmful interference which may be caused by these applications. ISM equipment operating in this band is subject to the provisions of No. 1815.

753A In respect of the radiodetermination-satellite service in the band 2483.5–2500 MHz, the provisions of No. 953 do not apply.

753F The use of the band 2483.5–2500 MHz by the mobile-satellite and the radiodetermination-satellite services is subject to the application of the coordination and notification procedures set forth in Resolution 46 (WARC-92). Coordination of space stations of the mobile-satellite and radiodetermination-satellite services with respect to terrestrial services is required only if the power flux-density produced at the Earth's surface exceeds the limits in No. 2566. In respect of assignments operating in this band, the provisions of Section II, paragraph 2.2 of Resolution 46 (WARC-92) shall also be applied to geostationary transmitting space stations with respect to terrestrial stations.

770 In the band 2700–2900 MHz, ground-based radars used for meteorological purposes are authorized to operate on a basis of equality with stations of the aeronautical radionavigation service.

775A In the bands 2900–3100 MHz and 9300–9500 MHz, the response from radar transponders shall not be capable of being confused with the response from radar beacons (racons) and shall not cause interference for ship or aeronautical radars in the radionavigation service, having regard, however, to No. 347 of these Regulations.

778 In making assignments to stations of other service, administrations are urged to take all practicable steps to protect the spectral line observations of the radio astronomy service from harmful interference in the bands 3260–3267 MHz, 3332–3339 MHz, 3345.8–3352.5 MHz and 4825–4835 MHz. Emissions from space or airborne stations can be particularly serious sources of interference to the radio astronomy service (see Nos. 343 and 344 and Article 36).

791 The standard frequency and time signal-satellite service may be authorized to use the frequency 4202 MHz for space-to-Earth transmissions and the frequency 6427 MHz for Earth-to-space transmissions. Such transmissions shall be confined within the limits of ± 2 MHz of these frequencies and shall be subject to agreement obtained under the procedure set forth in Article 14.

792A The use of the bands 4500–4800 MHz, 6725–7025 MHz, 10.7–10.95 GHz, 11.2–11.45 GHz and 12.75–13.25 GHz by the fixed-satellite service shall be in accordance with the provisions of Appendix 30B.

796 The band 5000–5250 MHz is to be used for the operation of the international standard system (microwave landing system) for precision approach and landing. The requirements of this system shall take precedence over other uses of this band.

797 The bands 5000–5250 MHz, and 15.4–15.7 GHz are also allocated to the fixed-satellite service and the inter-satellite service, for connection between one or more earth stations at specified fixed points on the Earth and space stations, when these services are used in conjunction with the aeronautical radionavigation and/or aeronautical mobile (R) service. Such use shall be subject to agreement obtained under the procedure set forth in Article 14.

797A Additional allocation: in the countries listed in Nos. 733B and 753C, and subject to agreement obtained under the procedure set forth in Article 14, and band 5150–5216 MHz is also allocated to the radiodetermination-satellite service (space-to-Earth) on a primary basis. In Region 2, the band is also allocated to the radiodetermination-satellite service (space-to-Earth) on a primary basis. In regions 1 and 3, except those countries listed in Nos. 733B and 753C, the band is also allocated to the radiodetermination-satellite service (space-to-Earth) on a secondary basis. The use by the radiodetermination-satellite service is limited to feeder links in conjunction with the radiodetermination-satellite service operating in the bands 1610–1626.5 MHz and/or 2483.5–2500 MHz. The total power flux-density at the Earth's surface shall in no case exceed -159 dBW/m$^2$ in any 4 kHz band for all angles of arrival.

799 The use of the band 5350–5470 MHz by the aeronautical radionavigational service is limited to airborne radars and associated airborne beacons.

802 Between 5600 MHz and 5650 MHz, ground-based radars used for meteorological purposes are authorized to operate on a basis of equality with stations of the maritime radionavigation service.

806 The band 5725–5875 MHz (centre frequency 5800 MHz) is designated for industrial, scientific and medical (ISM) applications. Radiocommunication services operating within this band must accept harmful interference which may be caused by these applications. ISM equipment operating in this band is subject to the provisions of No. 1815.

808 The band 5830–5850 MHz is also allocated to the amateur-satellite service (space-to-Earth) on a secondary basis.

809 In the band 6425–7075 MHz, passive microwave sensor measurements are carried out over the oceans. In the band 7075–7250 MHz, passive microwave sensor measurements are carried out. Administrations should bear in mind the needs of the earth exploration-satellite (passive) and space research (passive) services in their future planning of this band.

823 In the bands 8850–9000 MHz and 9200–9225 MHz, the maritime radionavigation service is limited to shore-based radars.

824A In the band 9200–9500 MHz, search and rescue transponders (SART) may be used, having due regard to the appropriate CCIR Recommendation (see also Article N 38).

825A In the band 9300–9320 MHz in the radionavigation service, the use of shipborne radars, other than those existing on 1 January 1976, is not permitted until 1 January 2001.

828 The band 9975–10025 MHz is also allocated to the meteorological-satellite service on a secondary basis for use by weather radars.

837 Different category of service: in Canada, Mexico and the United States, the allocation of the band 11.7–12.1 GHz to the fixed service is on a secondary basis (see No. 424).

839 The use of the bands 11.7–12.2 GHz by the fixed-satellite service in Region 2 and 12.2–12.7 GHz by the broadcasting-satellite service in Region 2 is limited to national and subregional systems. The use of the band 11.7–12.2 GHz by the fixed-satellite service in Region 2 is subject to previous agreement between the administrations concerned and those having services, operating or planned or operate in accordance with the table, which may be affected (see Articles 11, 13 and 14). For the use of the band 12.2–12.7 GHz by the broadcasting-satellite service in Region 2, see Article 15.

844 In Region 2, in the band 12.2–12.7 GHz, existing and future terrestrial radiocommunication services shall not cause harmful interference to the space services operating in conformity with the Broadcasting-Satellite Plan for Region 2 contained in Appendix 30 (Orb-85).

851 The use of the band 13.25–13.4 GHz by the aeronautical radionavigation service is limited to Doppler navigation aids.

855A (Proposed:) In the band 13.75–14 GHz, the e.i.r.p. of any emission from an earth station in the fixed-satellite service shall be at least 68 dBW, and should not exceed 85 dBW, with a minimum antenna diameter of 4.5 metres. In addition the e.i.r.p., averaged over one second, radiated by a station in the radiolocation or radionavigation services towards the geostationary-satellite orbit shall not exceed 59 dBW (see the latest version of Recommendation ITU-R S.1068 for additional information).

855B (Proposed:) In the band 13.75–14 GHz geostationary space stations in the space research service, for which information for advance publication has been received by the Bureau prior to 31 January 1992, shall operate on an equal basis with stations in the fixed-satellite service; after that date new geostationary space stations in the space research service will operate on a secondary basis. The e.i.r.p. density of emissions from any earth station in the fixed-satellite service shall not exceed 71 dBW per 6 MHz in the frequency range 13.772–13.778 GHz until those geostationary space stations in the space research service, for which information for advance publication has been received by the Bureau prior to 31 January 1992, cease to operate in this band. Automatic power control may be used to increase the e.i.r.p. density above 71 dBW to compensate for rain attenuation to the extent that the power flux density at the fixed-satellite space station does not exceed the value resulting from use of 71 dBW e.i.r.p. in clear sky conditions.

855C (Proposed:) Until 1 January 2000 (until 1 January 2001 for space borne precipitation radars), stations in the fixed-satellite service shall not cause harmful interference to non-geostationary space stations in the space research and Earth exploration satellite services. After those dates these non-geostationary space stations will operate on a secondary basis in relation to the fixed satellite service. When planning and coordinating networks in the fixed-satellite service in accordance with the provisions of Article 11 (S9) in the band 13.75–14 GHz, account should be taken of the information given in the latest version of Recommendation ITU-R S.1069.

862 In making assignments to stations of other services to which the band 14.47–14.5 GHz is allocated, administrations are urged to take all practicable steps to protect spectral line observations of the radio astronomy service from harmful interference. Emissions from space or airborne stations can be particularly serious sources of interference to the radio astronomy service (see Nos. 343 and 344 and Article 36).

870 The band 18.1–18.3 GHz is also allocated to the meteorological-satellite service (Earth-to-space) on a primary basis. Its use is limited to geostationary satellites and shall be in accordance with the provisions of No. 2578.

873A In order to facilitate interregional coordination between networks in the mobile-satellite and fixed-satellite services, carriers in the mobile-satellite service that are most susceptible to interference shall, to the extent practicable, be located in the higher parts of the bands 19.7–20.2 GHz and 29.5–30 GHz.

873B In the band 19.7–20.2 GHz and 29.5–30 GHz in Region 2, and in the bands 20.1–20.2 GHz and 29.9–30 GHz in Regions 1 and 3, networks which are both in the fixed-satellite service and in the mobile-satellite service may include links between earth stations at specified or unspecified points or while in motion, through one or more satellites for point-to-point and point-to-multipoint communications.

873C In the bands 19.7–20.2 GHz and 29.5–30 GHz, the provisions of No. 935 do not apply with respect to the mobile-satellite service.

873D The allocation to the mobile-satellite service is intended for use by networks which use narrow spot-beam antennas and other advanced technology at the space stations. Administrations operating systems in the mobile-satellite service in the band 19.7–20.1 GHz in Region 2 and in the band 20.1–20.2 GHz shall take all practicable steps to ensure the continued availability of these bands for administrations operating fixed and mobile systems in accordance with the provisions of No. 873.

873E The use of the bands 19.7–20.1 GHz and 29.5–29.9 GHz by the mobile-satellite service in Region 2 is limited to satellite networks which are both in the fixed-satellite service and in the mobile-satellite service as described in No. 873B.

874 In making assignments to stations of other services, administrations are urged to take all practicable steps to protect the spectral line observations of the radio astronomy service in the band 22.01–22.21 GHz from harmful

interference. Emissions from space or airborne stations can be particularly serious sources of interference to the radio astronomy service (see also Nos. 343 and 344 and Article 36).

875 In making assignments to stations of other services, administrations are urged to take all practicable steps to protect the radio astronomy service from harmful interference in the band 22.21–22.5 GHz. Emissions from space or airborne stations can be particularly serious sources of interference to the radio astronomy service (see also Nos. 343 and 344 and Article 36).

879 In making assignments to stations of other services, administrations are urged to take all practicable steps to protect the spectral line observations of the radio astronomy service in the bands 22.81–22.86 GHz and 23.07–23.12 GHz from harmful interference. Emissions from space or airborne stations can be particularly serious sources of interference to the radio astronomy service (see Nos. 343 and 344 and Article 36).

881 The band 24–24.25 GHz (centre frequency 24.125 GHz) is designated for industrial, scientific and medical (ISM) applications. Radiocommunication services operating within this band must accept harmful interference which may be caused by these applications. ISM equipment operating in this band is subject to the provisions of No. 1815.

882 The band 29.95–30 GHz may be used for space-to-space links in the earth exploration-satellite service for telemetry, tracking, and control purposes, on a secondary basis.

882E The inter-satellite service shall not claim protection from harmful interference from airport surface detection equipment stations of the radionavigation service.

886 In making assignments to stations of other services, administrations are urged to take all practicable steps to

protect the radio astronomy service from harmful interference in the band 31.2–31.3 GHz. Emissions from space or airborne stations can be particularly serious sources of interference to the radio astronomy service (see Nos. 343 and 344 and Article 36).

893 In designing systems for the inter-satellite and radionavigation services in the band 32–33 GHz, administrations shall take all necessary measures to prevent harmful interference between these two services, bearing in mind the safety aspects of the radionavigation service (see Recommendation 707).

897 Radars located on spacecraft may be operated on a primary basis in the band 35.5–35.6 GHz.

898 In making assignments to stations of other services, administrations are urged to take all practicable steps to protect the spectral line observations of 36.43–36.5 GHz from harmful interference. Emission from space or airborne stations can be particularly serious sources of interference to the radio astronomy service (see Nos. 343 and 344 and Article 36).

900 In making assignments to stations of other services to which the band 42.5–43.5 GHz is allocated, administrations are urged to take all practicable steps to protect the radio astronomy service from harmful interference, especially in the bands 42.77–42.87 GHz, 43.07–43.17 GHz, and 43.37–43.47 GHz, which are used for spectral line observations of silicon monoxide. Emissions from space or airborne stations can be particularly serious sources of interference to the radio astronomy service (see Nos. 343 and 344 and Article 36).

902 In the bands 43.5–47 GHz, 66–71 GHz, 95–100 GHz, 134–142 GHz, 190–200 GHz and 252–265 GHz, stations in the land mobile service may be operated subject to not causing harmful interference to the space radiocommunication services to which these bands are

allocated (see No. 435).

903 In the bands 43.5–47 GHz, 66–71 GHz, 95–100 GHz, 134–142 GHz, 190–200 GHz and 252–265 GHz, satellite links connecting land stations at specified fixed points are also authorized when used in conjunction with the mobile-satellite service or the radionavigation-satellite service.

904 The bands 48.94–49.04 GHz and 97.88–98.08 GHz are also allocated to the radio astronomy service on a primary basis for spectral line observations. In making assignments to stations of other services to which these bands are allocated, administrations are urged to take all practicable steps to protect the radio astronomy service from harmful interference. Emissions from space or airborne stations can be particularly serious sources of interference to the radio astronomy service (see Nos. 343 and 344 and Article 36).

909 In the bands 54.25–58.2 GHz 59–64 GHz, 116–134 GHz, 170–182 GHz and 185–190 GHz, stations in the aeronautical mobile service may be operated subject to not causing harmful interference to the inter-satellite service (see No. 435).

910 In the bands 59–64 GHz and 126–134 GHz, airborne radars in the radiolocation service may be operated subject to not causing harmful interference to the inter-satellite service (see No. 435).

911 The band 61–61.5 GHz (centre frequency 61.25 GHz) is designated for industrial, scientific and medical (ISM) applications. The use of this frequency band for ISM applications shall be subject to special authorization by the administrations concerned in agreement with other administrations whose radiocommunication services might be affected, In applying this provision administrations shall have due regard to the latest CCIR Recommendations.

912 In the band 78–79 GHz, radars located on space

stations may be operated on a primary basis in the earth exploration-satellite service and in the space research service.

913 In the band 84–86 GHz, stations in the fixed, mobile and broadcasting services shall not cause harmful interference to broadcasting-satellite stations operating in accordance with the decisions of the appropriate frequency assignment planning conference for the broadcasting-satellite service.

914 The band 93.07–93.27 GHz is also used by the radio astronomy service for spectral line observations. In making assignments to stations of the services to which this band is allocated, administrations are urged to take all practicable steps to protect radio astronomy observations from harmful interference. Emissions from space or airborne stations can be particularly serious sources of interference to the radio astronomy service (see Nos. 343 and 344 and Article 36).

915 The band 119.98–120.02 GHz is also allocated to the amateur service on a secondary basis.

916 The band 122–123 GHz (centre frequency 122.5 GHz) is designated for industrial, scientific and medical (ISM) applications. The use of this frequency band for ISM applications shall be subject to special authorization by the administration concerned in agreement with other administrations whose radiocommunication services might be affected. In applying this provision administrations shall have due regard to the latest CCIR Recommendations.

917 In the bands 140.69–140.98 GHz all emissions from airborne stations, and from space stations in the space-to-Earth direction, are prohibited.

918 The band 140.69–140.98 GHz, 144.68–144.98 GHz, 145.45–145.75 GHz and 146.82–147.12 GHz are also allocated to the radio astronomy service on a primary basis for spectral line observations. In making assignments to stations of other services to which the bands are allocated,

administrations are urged to take all practicable steps to protect the radio astronomy service (see Nos. 343 and 344 and Article 36).

919 The bands 150–151 GHz, 174.42–175.02 GHz, 177–177.4 GHz, 178.2–178.6 GHz, 181–181.46 GHz and 186.2–186.6 GHz are also allocated to the radio astronomy service on a secondary basis for spectral line observation. In making assignments to stations of other services to which these bands are allocated, administrations are urged to take all practicable steps to protect the radio astronomy service from harmful interference. Emissions from space or airborne stations can be particularly serious sources of interference to the radio astronomy service (see Nos. 343 and 344 and Article 36).

922 The band 244–246 GHz (centre frequency 245 GHz) is designated for industrial, scientific and medical (ISM) applications. The use of this frequency band for ISM applications shall be subject to special authorization by the administration concerned in agreement with other administrations whose radiocommunication services might be affected. In applying this provision administrations shall have due regard to the latest CCIR Recommendations.

923 The bands 250–251 GHz and 262.24–262.76 GHz are also allocated to the radio astronomy service on a primary basis for spectral line observations. In making assignments to stations of other services to which the band is allocated, administrations are urged to take all practicable steps to protect the radio astronomy service from harmful interference. Emissions from space or airborne stations can be particularly serious sources of interference to the radio astronomy service (see Nos. 343, 344 and Article 36).

924 The band 257.5–258 GHz is also allocated to the radio astronomy service on a secondary basis for spectral line observations. In making assignments to stations of other services to which the band is allocated,

administrations are urged to take all practicable steps to protect the radio astronomy service from harmful interference. Emissions from space or airborne stations can be particularly serious sources of interference to the radio astronomy service (see Nos. 343, 344 and Article 36).

926 In making assignments to stations of other services to which the band 265–275 GHz is allocated, administrations are urged to take all practicable steps to protect the radio astronomy service from harmful interference, especially in the bands 265.64–266.16 GHz, 267.34–267.86 GHz and 271.74–272.26 GHz, which are used for spectral line observations. Emissions from space or airborne stations can be particularly serious sources of interference to the radio astronomy service (see Nos. 343, 344 and Article 36).

927 The frequency band 275 GHz–400 GHz may be used by administrations for experimentation with, and development of, various active and passive services. In this band a need has been identified for the following spectral line measurements for passive services: (a) Radio astronomy service: 278–280 GHz and 343–348 GHz. (b) Space research service (passive) and earth exploration-satellite service (passive): 275–277 GHz, 300–302 GHz, 324–326 GHz, 345–347 GHz, 363–365 GHz and 379–381 GHz.

Future research in this largely unexplored spectral region may yield additional spectral lines and continuum bands of interest to the passive services. Administrations are urged to take all practicable steps to protect these passive services from harmful interference until the next competent world administrative radio conference.

# U.S. FOOTNOTES

These footnotes, each consisting of the letters US followed by one or more digits, denote stipulations applicable to both federal government and non-federal government stations. Only those U.S. footnotes that pertain to allocations in 30 MHz–300 GHz are shown.

US7 In the band 420–450 MHz and within the following areas, the peak envelope power output of a transmitter employed in the amateur service shall not exceed 50 watts, unless expressly authorized by the Commission after mutual agreement, on a case-by-case basis, between the Federal Communications Commission Engineer in Charge at the applicable district office and the military area frequency coordinator at the applicable military base. For areas (e) thru (j), the appropriate military coordinator is located at Peterson AFB, CO.

(a) Those portions of Texas and New Mexico bounded on the south by latitude 31° 45' North, on the east by 104° 00' West, on the north by latitude 34° 30' North, and on the west by longitude 107° 30' West;

(b) The entire State of Florida including the Key West area and the areas enclosed within a 200-mile radius of Patrick Air Force Base, Florida (latitude 28° 21' North, longitude 80° 43' West), and within a 200-mile radius of Eglin Air Force Base, Florida (latitude 30° 30' North, longitude 86° 30' West);

(c) The entire State of Arizona;

(d) Those portions of California and Nevada south of latitude 37° 10' North, and the areas enclosed within a 200 mile radius of the Pacific Missile Test Center, Point Mugu, California (latitude 34° 09' North, longitude 119° 11' West);

(e) In the State of Massachusetts within a 160-kilometer (100 mile) radius around locations at Otis Air Force Base, Massachusetts (latitude 41° 45' North, longitude 70° 32' West);

(f) In the State of California within a 240-kilometer (150 mile) radius around locations at Beale Air Force Base, California (latitude 39° 08' North, longitude 121° 26' West);

(g) In the State of Alaska within a 160 kilometer (100 mile) radius of Clear, Alaska (latitude 64 degrees, 17' north, longitude 149 degrees 10' west).

(h) In the State of North Dakota within a 160 kilometer (100 mile) radius of Concrete, North Dakota (latitude 48 degrees 43' north, longitude 97 degrees 54' west).

(i) In the States of Alabama, Florida, Georgia and South Carolina within a 200 kilometer (124 mile) radius of Warner Robins Air Force Base, Georgia (latitude 32° 38' North, longitude 83° 35' West).

(j) In the State of Texas within a 200 kilometer (124 mile) radius of Goodfellow Air Force Base, Texas (latitude 31° 25' North, longitude 100° 24' West).

US8 The use of the frequencies 170.475, 171.425, 171.575, and 172.275 MHz east of the Mississippi River, and 170.425, 170.575, 171.475, 172.225 and 172.375 MHz west of the Mississippi River may be authorized to fixed, land and mobile stations operated by non-Federal forest firefighting agencies. In addition, land stations and mobile stations operated by non-Federal conservation agencies, for mobile relay operation only, may be authorized to use the frequency 172.275 MHz east of the Mississippi River and the frequency 171.475 MHz west of the Mississippi River. The use of any of the foregoing nine frequencies shall be on the condition that no harmful interference will be caused to Government stations.

US10 The use of the frequencies 26.62, 143.75, 143.90 and 148.15 MHz may be authorized to Civil Air Patrol land stations and Civil Air Patrol mobile stations.

US11 The use of the frequencies 166.250 and 170.150 MHz may be authorized to non-Government remote pickup broadcast base and land mobile stations and to non-Government base, fixed and land mobile stations in the public safety radio services (the sum of the bandwidth of emission and tolerance is not to exceed 25 kHz, except that authorizations in existence as of December 20, 1974, using a larger bandwidth are permitted to continue in operation until December 20, 1979) in the continental United States (excluding Alaska) only, except within the area bounded on the west by the Mississippi River, on the north by the parallel of latitude 37° 30' N., and on the east and south by that arc of the circle with center at Springfield, Illinois, and radius equal to the airline distance between Springfield, Illinois, and Montgomery, Alabama, subtended between the foregoing west and north boundaries, on the condition that harmful interference will not be caused to Government stations present or future in the Government band 162–174 MHz. The use of these frequencies by remote pickup broadcast stations will not be authorized for locations within 150 miles of New York City; and use of these frequencies by the public safety radio services will not be authorized except for locations within 150 miles of New York City.

US13 For the specific purpose of transmitting hydrological and meteorological data in co-operation with agencies of the Federal Government, the following frequencies may be authorized to non-Government fixed stations on the condition that harmful interference will not be caused to Government stations.

MHz

169.425 169.525 170.300 171.075 171.850 406.125 412.625

169.450 170.225 170.325 171.100 171.875 406.175 412.675

169.475 170.250 171.025 171.125 171.900 409.675 412.725

169.500 170.275 171.050 171.825 171.925 409.725 412.775

Licensees holding a valid authorization on June 11, 1962, to operate on the frequencies 169.575, 170.375 or 171.975 MHz may continue to be authorized for such operations on the condition that harmful interference will not be caused to Government stations.

US18 Navigation aids in the US and possessions in the bands 9–14 kHz, 90–110 kHz, 190–415 kHz, 510–535 kHz, 2700–2900 MHz are normally operated by the U.S. Government. However, authorizations may be made by the FCC for non-government operation in these bands subject to the conclusion of appropriate arrangements between the FCC and the Government agencies concerned and upon special showing of need for service which the Government is not yet prepared to render.

US26 The bands 117.975–121.4125 MHz, 123.5875–128.8125 MHz and 132.0125–136.0 MHz are for air traffic control communications.

US28 The band 121.5875–121.9375 MHz is for use by aeronautical utility land and mobile stations, and for air traffic control communications.

US30 The band 121.9375–123.0875 MHz is available to FAA aircraft for communications pursuant to flight inspection functions in accordance with the Federal Aviation Act of 1958.

US31 Except as provided below the band 121.9375–123.0875 MHz is for use by private aircraft stations.

The frequencies 122.700, 122.725, 122.750, 122.800, 122.950, 122.975, 123.000, 123.050 and 123.075 MHz may be assigned to aeronautical advisory stations. In addition, at landing areas having a part-time or no airdrome control tower or FAA flight service station, these frequencies may be assigned on a secondary non-interference basis to aeronautical utility mobile stations, and may be used by FAA ground vehicles for safety related communications during inspections conducted at such landing areas.

The frequencies 122.850, 122.900 and 122.925 MHz may be assigned to aeronautical multicom stations. In addition, 122.850 MHz may be assigned on a secondary non-interference basis to aeronautical utility mobile stations. In case of 122.925 MHz, US213 applies.

Air carrier aircraft stations may use 122.000 and 122.050 MHz for communication with aeronautical stations of the Federal Aviation Administration and 122.700, 122.800, 122.900 and 123.000 MHz for communications with aeronautical stations pertaining to safety of flight with and in the vicinity of landing areas not served by a control tower.

Frequencies in the band 121.9375–122.6875 MHz may be used by aeronautical stations of the Federal Aviation Administration for communication with private aircraft stations only, except that 122.000 and 122.050 MHz may also be used for communication with air carrier aircraft stations concerning weather information.

US32 Except for the frequencies 123.3 and 123.5 MHz, which are not authorized for Government use, the band 123.1125–123.5875 MHz is available for FAA communications incident to flight test and inspection activities pertinent to aircraft and facility certification on a secondary non-interference basis.

US33 The band 123.1125–123.5875 MHz is for use by flight test and aviation instructional stations. The frequency

121.950 MHz is available for aviation instructional stations.

US41 The Government radiolocation service is permitted in the band 2450–2500 MHz on condition that harmful interference is not caused to non-Government services.

US44 The non-Government radiolocation service may be authorized in the band 2900–3100 MHz on the condition that no harmful interference is caused to Government services.

US48 The non-Government radiolocation service may be authorized in the bands 5350–5460 MHz and 9000–9200 MHz on the condition that it does not cause harmful interference to the aeronautical radionavigation service or to the Government radiolocation service.

US49 The non-Government radiolocation service may be authorized in the band 5460–5470 MHz on the condition that it does not cause harmful interference to the aeronautical or maritime radionavigation services or to the Government radiolocation service.

US50 The non-Government radiolocation service may be authorized in the band 5470–5600 MHz on the condition that it does not cause harmful interference to the maritime radionavigation service or to the Government radiolocation service.

US51 In the band 5600–5650 MHz and 9300–9500 MHz, the non-Government radiolocation service shall not cause harmful interference to the Government radiolocation service.

US53 In view of the fact that the band 13.25–13.4 GHz is allocated to doppler navigation aids, Government, and non-Government airborne doppler radars in the aeronautical radionavigation service are permitted in the band 8750–8850 MHz only on the condition that they must accept any interference that may be experienced from stations in the radiolocation service in the band 8500–10000 MHz.

US54 Temporarily, and until certain operations of the radiolocation service in the band 9000–9200 MHz can be transferred to other appropriate frequency bands, the aeronautical radionavigation service may, in certain geographical areas, be subject to receiving some degree of interference from the radiolocation service.

US58 In the band 10000–10500 MHz, pulsed emissions are prohibited, except for weather radars on board meteorological satellites in the band 10000–10025 MHz. The amateur service and the non-Government radiolocation service, which shall not cause harmful interference to the Government radiolocation service, are the only non-Government services permitted in this band. The non-Government radiolocation service is limited to survey operations as specified in footnote US108.

US59 The band 10.5–10.55 GHz is restricted to systems using type N0N (A0) emission with a power not to exceed 40 watts into the antenna.

US65 The use of the band 5460–5650 MHz by the maritime radionavigation service is limited to shipborne radars.

US66 The use of the band 9300–9500 MHz by the aeronautical radionavigation service is limited to airborne radars and associated airborne beacons. In addition, ground-based radar beacons in the aeronautical radionavigation service are permitted in the band 9300–9320 MHz on the condition that harmful interference is not caused to the maritime radionavigation service.

US67 The use of the band 9300–9500 MHz by the meteorological aids service is limited to ground-based radars. Radiolocation installations will be coordinated with the meteorological aids service and, insofar as practicable, will be adjusted to meet the requirements of the meteorological aids service.

US69 In the band 31.8–33.4 GHz, ground-based radionavigation aids are not permitted except where they operate in cooperation with airborne or shipborne radionavigation devices.

US70 The meteorological aids service allocation in the band 400.15–406.0 MHz does not preclude the operation therein of associated ground transmitters.

US71 In the band 9300–9320 MHz, low-powered maritime radionavigation stations shall be protected from harmful interference caused by the operation of land-based equipment.

US74 In the bands 25.55–25.67, 73.0–74.6, 406.1–410.0, 608–614, 1400–1427, 1660.5–1670.0, 2690–2700 and 4990–5000 MHz and in the bands 10.68–10.7, 15.35–15.4, 23.6–24.0, 31.3–31.5, 86–92, 105–116 and 217–231 GHz, the radio astronomy service shall be protected from extraband radiation only to the extent that such radiation exceeds the level which would be present if the offending station were operating in compliance with the technical standards or criteria applicable to the service in which it operates.

US77 Government stations may also be authorized:

(a) Port operations use on a simplex basis by coast and ship stations of the frequencies 156.6 and 156.7 MHz;

(b) Duplex port operations use of the frequency 157.0 MHz for ship stations and 161.6 MHz for coast stations;

(c) Inter-ship use of 156.3 MHz on a simplex basis; and

(d) Vessel traffic services under the control of the U.S. Coast Guard on a simplex basis by coast and ship stations on the frequencies 156.25, 156.55, 156.6 and 156.7 MHz.

(e) Navigational bridge-to-bridge and navigational communications on a simplex basis by coast and ship stations on the frequencies 156.375 and 156.65 MHz.

US78 In the mobile service, the frequencies between 1435 and 1535 MHz will be assigned for aeronautical telemetry and associated telecommand operations for flight testing of manned or unmanned aircraft and missiles, or their major components. Permissible usage includes telemetry associated with launching and reentry into the earth's atmosphere as well as any incidental orbiting prior to reentry of manned objects undergoing flight tests. The following frequencies are shared with flight telemetry mobile stations: 1444.5, 1453.5, 1501.5, 1515.5, 1524.5 and 1525.5 MHz.

US80 Government stations may use the frequency 122.9 MHz subject to the following conditions:

(a) All operations by Government stations shall be restricted to the purpose for which the frequency is authorized to non-Government stations, and shall be in accordance with the appropriate provisions of the Commission's Rules and Regulations, Part 87, Aviation Services;

(b) Use of the frequency is required for coordination of activities with Commission licensees operating on this frequency; and

(c) Government stations will not be authorized for operation at fixed locations.

US81 The band 38.0–38.25 MHz is used by both Government and non-government radio astronomy observatories. No new fixed or mobile assignments are to be made and Government stations in the band 38.0–38.25 MHz will be moved to other bands on a case-by-case basis, as required, to protect radio astronomy observations from harmful interference. As an exception, however, low powered military transportable and mobile stations used for tactical and training purposes will continue to use the band. To the extent practicable, the latter operations will be adjusted to relieve such interference as may be caused to radio astronomy observations. In the event of harmful interference from such local operations, radio astronomy observatories may contact local military commands directly, with a view to effecting relief. A list of military commands, areas of coordination, and points of contact for purposes of relieving interference may be obtained upon request from the Office of the Chief Engineer, Federal Communications Commission, Washington, D.C. 20554.

US87 The frequency 450 MHz, with maximum emission bandwidth of 500 kHz, may be used by Government and non-Government stations for space telecommand at specific locations, subject to such conditions as may be applied on a case-by-case basis.

US90 In the band 2025–2110 MHz Earth-to-space and space-to-space transmissions may be authorized in the space research and earth exploration-satellite services subject to such conditions as may be applied on a case-by case basis. Such transmissions shall not cause harmful interference to non-government stations operating in accordance with the Table of Frequency Allocations. All space-to-space transmissions reaching the Earth's surface shall adhere to a power flux density of between -144 and -154 dBW/m$^2$/4 kHz depending on the angle of arrival per ITU Radio Regulation 2557 and shall not cause harmful interference to the other space services.

US93 In the conterminous United States, the frequency 108.0 MHz may be authorized for use by VOR test facilities, the operation of which is not essential for the safety of life or property, subject to the condition that no interference is caused to the reception of FM broadcasting stations operating in the band 88–108 MHz. In the event that such interference does occur, the licensee or other agency authorized to operate the facility shall discontinue operation on 108 MHz and shall not resume operation until the interference has been eliminated or the complaint otherwise satisfied. VOR test facilities operating on 108 MHz will not be protected against interference caused by FM broadcasting stations operating in the band 88–108 MHz not shall the authorization of a VOR test facility on 108 MHz preclude the Commission from authorizing additional FM broadcasting stations.

US99 In the band 1668.4–1670.0 MHz, the meteorological aids service (radiosonde) will avoid operations to the maximum extent practicable. Whenever it is necessary to operate radiosondes in the band 1668.4–1670 MHz within the United States, notification of the operations shall be sent as far in advance as possible to the Electromagnetic Management Unit, National Science Foundation, Washington, D.C. 20550.

US102 In Alaska only, the frequency 122.1 MHz may also be used for air carrier air traffic control purposes at locations where other frequencies are not available to air carrier aircraft stations for air traffic control.

US106 The frequency 156.75 MHz is available for assignment to non-government and Government stations for environmental communications in accordance with an agreed plan.

US107 The frequency 156.8 MHz is the national distress, safety and calling frequency for the maritime mobile VHF radiotelephone service for use by Government and non-Government ship and coast stations. Guard bands of 156.7625–156.7875 and 156.8125–156.8375 MHz are maintained.

US108 Within the bands 3300–3500 MHz and 10000–10500 MHz, survey operations, using transmitters with a peak power not to exceed five watts into the antenna, may be authorized for Government and non-Government use on a secondary basis to other Government radiolocation operations.

US110 (Proposed:) In the frequency bands 3100–3300 MHz, 3500–3700 MHz, 5250–5350 MHz, 8500–9000 MHz, 9200–9300 MHz, 9500–10000 MHz, 13.4–14.0 GHz, 15.7–17.3 GHz, 24.05–24.25 GHz and 33.4–36.0 GHz, the non-Government radiolocation service shall be secondary to the Government radiolocation service and to airborne doppler radars at 8800 MHz, and shall provide protection to airport surface detection equipment (ASDE) operating between 15.7–16.2 GHz.

US111 In the band 1990–2120 MHz, Government space research earth stations may be authorized to use specific frequencies at specific locations for Earth-to-space transmissions. Such authorizations shall be secondary to non-government use of this band and subject to such other conditions as may be applied on a case-by-case basis.

Corpus Christi, Tex., 27° 39' N., 097° 23' W.

Fairbanks, Alaska, 64° 59' N., 147° 53' W.

Goldstone, Calif., 35° 18' N., 116° 54' W.

Greenbelt, Md., 39° 00' N., 076° 50' W.

Guam, Mariana Is., 13° 19' N., 144° 44' E.

Kauai, Hawaii, 22° 08' N., 159° 40' W.

Merritt Is., Fla., 28° 29' N., 080° 35' W.

Rosman, N.C., 35° 12' N., 082° 52' W.

Wallops Is., Va., 37° 57' N., 075° 20' W.

US112 The frequency 123.1 MHz is for search and rescue communications. This frequency may be assigned for air traffic control communications at special aeronautical events on the condition that no harmful interference is caused to search and rescue communications during any period of search and rescue operations in the locale involved.

US116 In the bands 890–902 MHz and 935–941 MHz, no new assignments are to be made to Government radio stations after July 10, 1970 except on case-by case basis, to experimental stations and to additional stations of existing networks in Alaska. Government assignments existing prior to July 10 1970 to stations in Alaska may be continued. All other existing Government assignments shall be on a secondary basis to stations in the non-Government land mobile service and shall be subject to adjustment or removal from the bands 890–902 MHz, 928–932 MHz and 935–941 MHz at the request of the FCC.

US117 In the band 406.1–410 MHz, all new authorizations will be limited to a maximum 7 watts per kHz of necessary bandwidth; existing authorizations as of November 30, 1970 exceeding this power are permitted to continue in use.

New authorizations in this band stations, other than mobile stations, within the following areas are subject to prior coordination by the applicant through the Electromagnetic Spectrum Management Unit, National Science Foundation, Washington, D.C. 20550, (202-357-9696):

Arecibo Observatory:

Rectangle between latitudes 17° 30' N. and 19° 00' N. and between longitudes 65° 10' W. and 68° 00' W.

Owens Valley Radio Observatory:

Two contiguous rectangles, one between latitudes 36° N. and 37° N. and longitudes 117° 40' W. and 118° 30' W. and the second between latitudes 37° N. and 38° N. and longitudes 118° W. and 118° 50' W.

Sagamore Hill Radio Observatory:

Rectangle between latitudes 42° 10' N. and 43° 00' N. and longitudes 70° 31' W. and 71° 31' W.

Table Mountain Solar Observatory (NOAA), Boulder, Colorado (407–409 MHz only):

Rectangle between latitudes 39° 30' N. and 40° 30' N. and longitudes 104° 30' W. and 106° 00' W. or the Continental Divide whichever is farther east.

The non-Government use of this band is limited to the radio astronomy service and as provided by footnote US13.

US201 In the band 460–470 MHz, space stations in the earth exploration-satellite service may be authorized for space-to-Earth transmissions on a secondary basis with respect to the fixed and mobile services. When operating in the meteorological-satellite service, such stations shall be protected from harmful interference from other applications of the earth exploration-satellite service. The power flux produced at the Earth's surface by any space station in this band shall not exceed -152 dBW/$m^2$/4 kHz.

US203 Radio astronomy observations of the formaldehyde line frequencies 4825–4835 MHz and 14.470–14.500 GHz may be made at certain radio astronomy observatories as indicated below:

| Bands to be Observed | | |
|---|---|---|
| 4 GHz | 14 GHz | Observatory |
| X | | National Astronomy and Ionosphere Center, Arecibo, Puerto Rico |
| X | X | National Radio Astronomy Observatory, Green Bank, W. Va. |
| X | X | National Radio Astronomy Observatory, Socorro, New Mexico |
| X | X | Hat Creek Observatory (U of Calif.), Hat Creek, Cal. |

| | | |
|---|---|---|
| X | X | Haystack Radio Observatory (MIT-Lincoln Lab), Tyngsboro, Mass. |
| X | X | Owens Valley Radio Observatory (Cal. Tech.), Big Pine, Cal. |
| | X | Five College Radio Astronomy Observatory, Quabbin Reservoir (near Amherst), Massachusetts |

Every practicable effort will be made to avoid the assignment of frequencies to stations in the fixed or mobile services in these bands. Should such assignments result in harmful interference to these observations, the situation will be remedied to the extent practicable.

US205 Tropospheric scatter systems are prohibited in the band 2500–2690 MHz.

US208 Planning and use of the band 1559–1626.5 MHz necessitate the development of technical and/or operational sharing criteria to ensure the maximum degree of electromagnetic compatibility with existing and planned systems within the band.

US209 The use of frequencies 460.6625, 460.6875, 460.7125, 460.7375, 460.7625, 460.7875, 460.8125, 460.8375, 460.8625, 465.6625, 465.6875, 465.7125, 465.7375, 465.7625, 465.7875, 465.8125, 465.8375, and 465.8625 MHz may be authorized, with 100 mW or less output power, to Government and non-government radio stations for one-way, non-voice bio-medical telemetry operations in hospitals, or medical or convalescent centers.

US210 Use of frequencies in the bands 40.66–40.70 and 216–220 MHz may be authorized to Government and non-Government stations on a secondary basis for the tracking of, and telemetering of scientific data from, ocean buoys and wildlife. Airborne wildlife telemetry in the 216–220 MHz band will be limited to the 216.0–216.1 MHz portion of the band. Operation in these two bands is subject to the technical standards specified in: (a) Section 8.2.42 of the NTIA Manual for Government use, or (b) Section 5.108 of the Commission's Rules for non-Government use.

US211 In the bands 1670–1690, 5000–5250 MHz and 10.7–11.7, 15.1365–15.35, 15.4–15.7, 22.5–22.55, 24–24.05, 31.0–31.3, 31.8–32.0, 40.5–42.5, 84–86, 102–105, 116–126, 151–164, 176.5–182, 185–190, 231–235, 252–265 GHz, applicants for airborne or space station assignments are urged to take all practicable steps to protect radio astronomy observations in the adjacent lands from harmful interference; however, US74 applies.

US213 The frequency 122.925 MHz is for use only for communications with or between aircraft when coordinating natural resources programs of Federal or State natural resources, agencies, including forestry management and fire suppression, fish and game management and protection and environmental monitoring and protection.

US214 The frequency 157.1 MHz is the primary frequency for liaison communications between ship stations and stations of the United States Coast Guard.

US215 Emissions from microwave ovens manufactured on and after January 1, 1980, for operation on the frequency 915 MHz must be confined within the band 902–928 MHz. Emissions from microwave ovens manufactured prior to January 1, 1980, for operation on the frequency 915 MHz must be confined within the band 902–940 MHz. Radiocommunications services operating in the band 928–940 MHz must accept any harmful interference from the operation of microwave ovens manufactured before January 1, 1980.

US216 The frequencies 150.775 and 150.790, and the bands 152–152.0150, 163.2375–163.2625, 462.9375–463.1875 and 467.9375–468.1875 MHz are authorized for Government/non-Government operations in medical radio communications systems.

US217 Pulse-ranging radiolocation systems may be authorized for Government and non-Government use in the 420–450 MHz band along the shorelines of Alaska and the contiguous 48 states. Spread spectrum radiolocation systems may be authorized in the 420–435 MHz portion of the band for operation within the contiguous 48 States and Alaska. Authorizations will be granted on a case-by-case basis; however, operations proposed to be located within the zones set forth in US228 should not expect to be accommodated. All stations operating in accordance with this provision will be secondary to stations operating in accordance with the Table of Frequency Allocations.

US218 The band 902–928 MHz is available for Location and Monitoring Service (LMS) systems subject to not causing harmful interference to the operation of all Government stations authorized in these bands. These systems must tolerate interference from the operation of industrial, scientific, and medical (ISM) devices and the operation of Government stations authorized in these bands.

US219 In the band 2025–2110 MHz Government Earth Resources Satellite Earth Stations in the Earth Exploration-Satellite Service may be authorized to use the frequency 2106.4 MHz for Earth-to-space transmission for tracking, telemetry, and telecommand at the sites listed below. Such transmissions shall not cause harmful interference to non-Government operations:

Sioux Falls, S.D., 43° 32' 03.1" N. 96° 45' 42.8" W.

Fairbanks, Alaska, 64° 58' 36.6" N., 147° 30' 54.2" W.

US220 The frequencies 36.25 and 41.71 MHz may be authorized to Government stations and non-Government stations in the petroleum radio service, for oil spill containment and cleanup operations. The use of these frequencies for oil spill containment or cleanup operations is limited to the inland and coastal waterway regions.

US222 In the band 2025–2035 MHz Geostationary Operational Environmental Satellite Earth stations in the Space Research and Earth Exploration-Satellite Services may be authorized on a coequal basis for Earth-to-space transmissions for tracking, telemetry, and telecommand at the sites listed below:

Wallops Is., Va. 37° 50' 48" N., 75° 27' 33" W.

Seattle, Wash. 47° 34' 15" N., 122° 33' 10" W.

Honolulu, Hawaii 21° 21' 12" N., 157° 52' 36" W.

US223 Within 75 miles of the United States/Canada border on the Great Lakes, the St. Lawrence Seaway, and the Puget Sound and the Strait of Juan de Fuca and its approaches, use of coast transmit frequency 162.025 MHz and ship station transmit frequency 157.425 MHz (VHF maritime mobile service Channel 88) may be authorized for use by the maritime service for public correspondence.

US224 Government systems utilizing spread spectrum techniques for terrestrial communication, navigation and identification may be authorized to operate in the band 960–1215 MHz on the condition that harmful interference will not be caused to the aeronautical radionavigation service. These systems will be handled on a case-by-case basis. Such systems shall be subject to a review at the national level for operational requirements and electromagnetic compatibility prior to development, procurement or modification.

US228 Applicants for operation in the band 420 to 450 MHz under the provisions of US217 should not expect to be accommodated if their area of service is within the following geographic areas:

(a) Those portions of Texas and New Mexico bounded on the south by latitude 31° 45' North, on the east by longitude 104° 00' West, on the north by latitude 34° 30' North, and on the West by longitude 107° 30' West.

(b) In the State of Massachusetts within a 160 kilometers (100 miles) radius around the locations of Otis Air Force Base, Massachusetts (latitude 41° 45' North, longitude 70° 32' West).

(c) In the State of California within a 240 kilometer (150 mile) radius of Beale Air Force Base, California (latitude 39° 08' North, longitude 121° 26' West).

(d) In the State of Alaska, within a 160 kilometer (100 mile) radius of Clear, Alaska (latitude 64° 17' North, longitude 149° 10' West).

(e) In the State of North Dakota, within a 160 kilometer (100 mile) radius of Concrete, North Dakota (latitude 48° 43' North, longitude 97° 54' West).

(f) Those portions of Texas and New Mexico bounded on the south by latitude 31° 45' North, on the east by longitude 104° 00' West, on the north by latitude 34° 30' North, and on the West by longitude 107° 30' West.

(g) In the state of Alaska within a 160 kilometer (100 mile) radius of Clear, Alaska (latitude 64 degrees 17' North, longitude 149 degrees 10' West).

(h) In the state of North Dakota within a 160 kilometer (100 mile) radius of Concrete, North Dakota (latitude 48 degrees 43' North, longitude 97 degrees 54' West).

(i) In the States of Alabama, Florida, Georgia and South Carolina within a 200 kilometer (124 mile) radius of Warner Robins Air Force Base, Georgia (latitude 32° 38' North, longitude 83° 35' West).

(j) In the State of Texas within a 200 kilometer (124 mile) radius of Goodfellow Air Force Base, Texas (latitude 31° 25' North, longitude 100° 24' West).

US229 Assignments to stations in the fixed and mobile services may be made on the condition that no harmful interference is caused to the Navy SPASUR system currently operating in the southern United States in the frequency band 216.88–217.08 MHz.

US230 Non-government land mobile service is allocated on a primary basis in the bands 422.1875–425.4875 and 427.1875–429.9875 MHz within 50 statute miles of Detroit, MI, and Cleveland, OH, and in the bands 423.8125–425.4875 and 428.8125–429.9875 MHz within 50 statute miles of Buffalo, NY.

US244 The band 136.000–137.000 MHz is allocated to the non-Government aeronautical mobile (R) service on a primary basis, and is subject to pertinent international treaties and agreements. The frequencies 136.000 MHz, 136.025 MHz, 136.050 MHz, 136.075 MHz, 136.125 MHz, 136.150 MHz, 136.175 MHz, 136.225 MHz, 136.250 MHz, 136.300 MHz, 136.325 MHz, 136.350 MHz, 136.400 MHz, 136.425 MHz and 136.450 MHz are available on a shared basis to the Federal Aviation Administration for air traffic control purposes, such as automatic weather observation services (AWOS), automatic terminal information services (ATIS) and airport control tower communications. Stations licensed prior to January 2, 1990, using the 136–137 MHz band for space operation (space-to-Earth), meteorological-satellite service (space-to-Earth) and the space research service (space-to-Earth) may continue to use this band on a secondary basis to aeronautical mobile (R) service stations. No new assignments will be made to stations in the above space services.

US245 The fixed-satellite service is limited to international intercontinental systems and subject to case-by-case electromagnetic compatibility analysis.

US246 No stations will be authorized to transmit in the bands 608–614 MHz, 1400–1427 MHz, 1660.5–1668.4 MHz, 2690–2700 MHz, 4990–5000 MHz, 10.68–10.70 GHz, 15.35–15.40 GHz, 23.6–24.0 GHz, 31.3–31.8 GHz, 51.4–54.25 GHz, 58.2–59.0 GHz, 64–65 GHz, 86–92 GHz,

100–102 GHz, 105–116 GHz, 164–168 GHz, 182–185 GHz and 217–231 GHz.

US251 The band 12.75–13.25 GHz is also allocated to the space research (deep space) (space-to-Earth) service for reception only at Goldstone, California, 35° 18' N. 116° 54' W.

US252 The bands 2110–2120, 7145–7190 MHz, and 34.2–34.7 GHz are also allocated for Earth-to-space transmissions in the space research service, limited to deep space communications at Goldstone, California.

US253 In the band 2300–2310 MHz, the fixed and mobile services shall not cause harmful interference to the amateur service.

US254 In the band 18.6–18.8 GHz the fixed and mobile services shall be limited to a maximum equivalent isotropically radiated power of +35 dBW and the power delivered to the antenna shall not exceed -3 dBW.

US255 In the band 18.6–18.8 GHz the fixed satellite service shall be limited to a power flux density at the Earth's surface of -101 dbW/M$^2$ in a 200 MHz band for all angles of arrival.

US256 Radio astronomy observations may be made in the band 1718.8–1722.2 MHz on an unprotected basis. Agencies providing other services in this band in the geographic areas listed below should bear in mind that their operations may affect those observations, and those agencies are encouraged to minimize potential interference to the observations in so far as it is practicable.

National Astronomy and Ionosphere Center, Arecibo, Puerto Rico:

Rectangle between latitudes 17° 30' N. and 19° 00' N. and between longitudes 65° 10' W. and 68° 00' W.

Haystack Radio Observatory, Tyngsboro, Massachusetts:

Rectangle between latitudes 41° 00' N. and 43° 00' N. and between longitudes 71° 00' W. and 73° 00' W.

National Radio Astronomy Observatory, Green Bank, West Virginia:

Rectangle between latitudes 37° 00' N. and 39° 15' N. and longitudes 78° 30' W. and 80° 30' W.

National Radio Astronomy Observatory, Socorro, New Mexico:

Rectangle between latitudes 32° 30' N. and 35° 30' N. and between longitudes 106° 00' W. and 109° 00' W.

Owens Valley Radio Observatory, Big Pine, California:

Two contiguous rectangles, one between latitudes 36° 00' N. and 37° 00' N. and between longitudes 117° 40' W. and 118° 30' W. and the second between latitudes 37° 00' N. and 38° 00' N. and longitudes 118° 00' W and 118° 50' W.

Hat Creek Observatory, Hat Creek, California:

Rectangles between latitudes 40° 00' N. and 42° 00' N. and between longitudes 120° 15' W. and 122° 15' W.

US257 Radio astronomy observations may be made in the 4950–4990 MHz band at certain Radio Astronomy Observatories indicated below:

Hat Creek Observatory, Hat Creek, California:

Rectangle between latitudes 40° 00' N. and 42° 00' N. and between longitudes 120° 15' W. and 122° 15' W.

Owens Valley Radio Observatory, Pine, California:

Two contiguous rectangles, Big one between latitudes 36° 00' N. and 37° 00' N. and longitudes 117° 40' W. and 118° 30' W. and the second between latitudes 37° 00' N. and 38° 00' N. and longitudes 118° 00' W. and 118° 50' W.

Haystack Radio Observatory, Tyngsboro, Massachusetts:

Rectangle between latitudes 41° 00' N. and 43° 00' N. and between longitudes 71° 00' W. and 73° 00' W.

National Astronomy and Ionosphere Center, Arecibo, Puerto Rico:

Rectangle between latitudes 17° 30' N. and 19° 00' N. and between longitudes 65° 10' W. and 68° 00' W.

National Radio Astronomy Observatory, Socorro, New Mexico:

Rectangle between latitudes 32° 30' N. and 35° 30' N. and longitudes 106° 00' W. and 109° 00' W.

National Radio Astronomy Observatory, Green Bank, West Virginia:

Rectangle between latitudes 37° 30' N. and 39° 15' N. and longitudes 78° 30' W. and 80° 30' W.

Every practicable effort will be made to avoid the assignment of frequencies in the band 4950–4990 MHz to stations in the fixed and mobile services within the geographic areas given above. In addition, every practicable effort will be made to avoid the assignment of frequencies in this band to stations in the aeronautical mobile service which operate outside of those geographic areas, but which may cause harmful interference to the listed observatories. Should such assignments result in harmful interference to these observatories, the situation will be remedied to the extent practicable.

US258 In the band 8025–8400 MHz, the non-Government earth exploration-satellite service (space-to-Earth) is allocated on a primary basis. Authorizations are subject to a case-by-case electromagnetic compatibility analysis.

US259 Stations in the radiolocation service in the band 17.3–17.7 GHz, shall be restricted to operating powers of less than 51 dBW eirp after feeder link stations for the

US259 Stations in the radiolocation service in the band 17.3–17.7 GHz, shall be restricted to operating powers of less than 51 dBW eirp after feeder link stations for the broadcasting-satellite service are authorized and brought into use.

US260 Aeronautical mobile communications which are an integral part of aeronautical radionavigation systems may be satisfied in the bands 1559–1626.5 MHz, 5000–5250 MHz and 15.4–15.7 GHz.

US261 The use of the band 4200–4400 MHz by the aeronautical radionavigation service is reserved exclusively for airborne radio altimeters. Experimental stations will not be authorized to develop equipment for operational use in this band other than equipment related to altimeter stations. However, passive sensing in the earth exploration-satellite and space research services may be authorized in this band on a secondary basis (no protection is provided from the radio altimeters).

US262 The band 31.8–32.3 GHz is also allocated for space-to-Earth transmissions in the space research service, limited to deep space communications at Goldstone, California.

US263 In the frequency bands 21.2–21.4, 22.21–22.5, 36–37, 50.2–50.4, 54.25–58.2, 116–126, 150–151, 174.5–176.5, 200–202 and 235–238 GHz, the space research and earth exploration-satellite services shall not receive protection from the fixed and mobile services operating in accordance with the Table of Frequency Allocations.

US264 In the band 48.94–49.04 GHz, airborne stations shall not be authorized.

US265 In the band 10.6–10.68 GHz, the fixed service shall be limited to a maximum equivalent isotropically radiated power of 40 dBW and the power delivered to the antenna shall not exceed -3dBW per 250 kHz.

US266 Licensees in the public safety radio services holding a valid authorization on June 30, 1958, to operate in the frequency band 156.27–157.47 MHz or on the frequencies 161.85, 161.91 or 161.97 MHz may, upon proper application, continue to be authorized for such operation, including expansion of existing systems, until such time as harmful interference is caused to the operation of any authorized station other than those licensed in the public safety radio service.

US267 In the band 902–928 MHz, amateur radio stations shall not operate within the States of Colorado and Wyoming, bounded by the area of: latitude 39°N. to 42°N. and longitude 103°W. to 108°W.

US268 The bands 890–902 MHz and 928–942 MHz are also allocated to the radiolocation service for Government ship stations (off-shore ocean areas) on the condition that harmful interference is not caused to non-Government land mobile stations. The provisions of footnote US116 apply.

US269 In the band 2500–2690 MHz, applicants for space station assignments are urged to take all practicable steps to protect radio astronomy observations in the adjacent band, 2690–2700 MHz, from harmful interference. Further, all applicants are urged to coordinate their proposed system through the Electromagnetic Management Unit, National Science Foundation, Washington, D.C. 20550, prior to system development.

US270 The band 72.77–72.91 GHz is also allocated to the radio astronomy service. Applicants for frequency assignments in this band are urged to take all practicable steps to protect radio astronomy observations from harmful interference.

US271 The use of the band 17.3–17.8 GHz by the fixed-satellite service (Earth-to-space) is limited to feeder links for broadcasting-satellite service.

US274 In the 216–220 MHz band fixed, aeronautical mobile and land mobile stations are limited to telemetering and associated telecommand operations.

US275 The band 902–928 MHz is allocated on a secondary basis to the amateur service subject to not causing harmful interference to the operations of Government stations authorized in this band or to Location and Monitoring Service (LMS) systems. Stations in the Amateur service must tolerate any interference from the operations of industrial, scientific and medical (ISM) devices, LMS systems and the operations of Government stations authorized in this band. Further, the Amateur Service is prohibited in those portions of Texas and New Mexico bounded on the south by latitude 31° 41' North, on the east by longitude 104° 11' West, on the north by latitude 34° 30' North, and on the west by longitude 107° 30' West; in addition, outside this area but within 150 miles of these boundaries of White Sands Missile Range the service is restricted to a maximum transmitter peak envelope power output of 50 watts.

US276 Except as otherwise provided for herein, use of the band 2310–2390 MHz by the mobile service is limited to aeronautical telemetering and associated telecommand operations for flight testing of manned or unmanned aircraft, missiles, or major components thereof.

The following six frequencies are shared on a co-equal basis for telemetering and associated telecommand operations of expendable launch vehicles whether or not such operations involve flight testing: 2312.5, 2332.5, 2352.5, 2364.5, 2370.5 and 2382.5 MHz. All other mobile telemetering uses shall be secondary to the above uses.

US277 The band 10.6–10.68 GHz is also allocated on a primary basis to the radio astronomy service. However, the radio astronomy service shall not receive protection from stations in the fixed service which are licensed to operate in the one hundred most populous urbanized areas as defined by the U.S. Census Bureau. The following radio astronomy sites have been coordinated for observations in this band:

National Radio Astronomy Observatory, Green Bank, West Virginia; (38° 26' 08" N.; 79° 49' 42" W.) National Radio Astronomy Observatory, Socorro, New Mexico; (34° 04' 43" N.; 107° 37' 04" W.), Harvard Radio Astronomy Station, Fort Davis, Texas; (30° 38' 08" N.; 103° 56' 42" W.), Hat Creek Observatory, Hat Creek, California; (40° 49' 03" N.; 121° 28' 24" W.), Owens Valley Radio Observatory, Big Pine, California; (37° 13' 54" N.; 118° 17' 36" W.), Naval Research Laboratory, Maryland Point, Maryland (38° 22' 26" N.; 77° 14' 00" W.).

US278 In the 22.55–23.55 and 32–33 GHz bands non-geostationary intersatellite links may operate on a secondary basis to geostationary intersatellite links.

US287 The band 14–14.5 GHz is also allocated to the non-Government land mobile-satellite service (Earth-to-space) on a secondary basis.

US291 Television pickup stations in the mobile service may be authorized to use frequencies in the band 38.6–40 GHz on a secondary basis to stations operating in accordance with the Table of Frequency Allocations.

US292 In the band 14.0–14.2 GHz stations in the radionavigation service shall operate on a secondary basis to the fixed-satellite service.

US297 The bands 47.2–49.2 GHz and 74.0–75.5 GHz are also available for feeder links for the broadcasting-satellite service.

US300 The frequencies 169.445, 169.505, 170.245, 170.305, 171.045, 171.105, 171.845 and 171.905 MHz are available for wireless microphone operations on a secondary basis to Government and non-Government operations.

US300 The frequencies 169.445, 169.505, 170.245, 170.305, 171.045, 171.105, 171.845 and 171.905 MHz are available for wireless microphone operations on a secondary basis to Government and non-Government operations.

US301 Except as provided in US302, broadcast auxiliary stations licensed as of November 21, 1984, to operate in the band 942–944 MHz may continue to operate on a co-equal primary basis to other stations and services operating in the band in accordance with the Table of Frequency Allocations.

US302 The band 942–944 MHz in Puerto Rico is allocated as an alternative allocation to the fixed service for broadcast auxiliary stations only.

US303 In the band 2285–2290 MHz, non-Government space stations in the space research, space operations and earth exploration-satellite services may be authorized to transmit to the Tracking and Data Relay Satellite System subject to such conditions as may be applied on a case-by-case basis. Such transmissions shall not cause harmful interference to authorized Government stations. The power flux density at the Earth's surface from such non-government stations shall not exceed -144 to -154 $dBW/m^2/4$ kHz, depending on angle of arrival, in accordance with ITU Radio Regulation 2557.

US307 The sub-band 5150–5216 MHz is also allocated for space-to-Earth transmissions in the fixed satellite service for feeder links in conjunction with the radiodetermination satellite service operating in the bands 1610–1626.5 MHz and 2483.5–2500 MHz. The total power flux density at the Earth's surface shall in no case exceed -159 dBW/m per 4 kHz for all angles of arrival.

US308 In the frequency bands 1549.5–1558.5 MHz and 1651–1660 MHz, the Aeronautical-Mobile-Satellite (R) requirements that cannot be accommodated in the 1545–

safety-related communications in the mobile-satellite service.

US309 Transmissions in the bands 1545–1559 MHz from terrestrial aeronautical stations directly to aircraft stations, or between aircraft stations, in the aeronautical mobile (R) service are also authorized when such transmissions are used to extend or supplement the satellite-to-aircraft links. Transmissions in the band 1646.5–1660.5 MHz from aircraft stations in the aeronautical mobile (R) service directly to terrestrial aeronautical stations, or between aircraft stations, are also authorized when such transmissions are used to extend or supplement the aircraft-to-satellite links.

US310 In the band 14.896–15.121 GHz, non-Government space stations in the space research service may be authorized on a secondary basis to transmit to Tracking and Data Relay Satellites subject to such conditions as may be applied on a case-by-case basis. Such transmissions shall not cause harmful interference to authorized Government stations. The power flux density at the Earth's surface from such non-Government stations shall not exceed -138 to -148 dBW/m$^2$/4 kHz, depending on the angle of arrival, in accordance with CCIR Recommendation 510-1.

US311 Radio astronomy observations may be made in the 1350–1400 MHz band on an unprotected basis at certain Radio Astronomy Observatories indicated below:

National Astronomy and Ionosphere Center, Arecibo, Puerto Rico:

Rectangle between latitudes 17° 30' N and 19° 00' N and between longitudes 65° 10' W and 68° 00' W.

National Radio Astronomy Observatory, Socorro, New Mexico:

Rectangle between latitudes 32° 30' N and 35° 30' N and between longitudes 106° 00' W and 109° 00' W.

National Radio Astronomy Observatory, Green Bank, West Virginia:

Rectangle between latitudes 37° 30' N and 39° 15' N and between longitudes 78° 30' W and 80° 30' W.

National Radio Astronomy Observatory: 50 mile radius circles centered on:

| Very long baseline array stations | Latitude (North) | Longitude (West) |
|---|---|---|
| Pie Town, NM | 34° 18’ | 108° 07’ |
| Kitt Peak, AZ | 31° 57’ | 111° 37’ |
| Los Alamos, NM | 35° 47’ | 106° 15’ |
| Fort Davis, TX | 30° 38’ | 103° 57’ |
| North Liberty, IA | 41° 46’ | 91° 34’ |
| Brewster, WA | 48° 08’ | 119° 41’ |
| Owens Valley, CA | 37° 14’ | 118° 17’ |
| Saint Croix, VI | 17° 46’ | 64° 35’ |
| Mauna Kea, HI | 19° 49’ | 155° 28’ |
| Hancock, NH | 42° 56’ | 71° 59’ |

Every practicable effort will be made to avoid the assignment of frequencies in the band 1350–1400 MHz to stations in the fixed and mobile services which could interfere with radio astronomy observations within the geographic areas.

US312 The frequency 173.075 MHz may also be authorized on a primary basis to non-Government stations in the Police Radio Service (with a maximum authorized bandwidth of 20 kHz) for stolen vehicle recovery systems.

US315 In the frequency bands 1530–1544 MHz and 1626.5–1645.5 MHz maritime mobile-satellite distress and safety communications, e.g., GMDSS, shall have priority

US312 The frequency 173.075 MHz may also be authorized on a primary basis to non-Government stations in the Police Radio Service (with a maximum authorized bandwidth of 20 kHz) for stolen vehicle recovery systems.

US315 In the frequency bands 1530–1544 MHz and 1626.5–1645.5 MHz maritime mobile-satellite distress and safety communications, e.g., GMDSS, shall have priority access with real-time preemptive capability in the mobile-satellite service. Communications of mobile-satellite system stations not participating in the GMDSS shall operate on a secondary basis to distress and safety communications of stations operating in the GMDSS. Account shall be taken of the priority of safety-related communications in the mobile-satellite service.

US316 The band 2900–3000 MHz is also allocated on a primary basis to the Meteorological Aids Service. Operations in this service are limited to Government Next Generation Weather Radar (NEXRAD) systems where accommodation in the 2700–2900 MHz band is not technically practical and are subject to coordination with existing authorized stations.

US317 The band 218.0–219.0 MHz is allocated on a primary basis to the Interactive Video and Data operations.

US318 Until January 1, 2000, the use of the 137–138 MHz band by the mobile-satellite service will be secondary to Government satellite operations in the subbands: 137.333–137.367, 137.485–137.515, 137.605–137.635 and 137.753–137.787 MHz.

US319 In the 137–138, 148–149.9, 149.9–150.05, 399.9–400.05, 400.15–401, 1610–1626.5, and 2483.5–2500 MHz bands, Government stations in the mobile-satellite service shall be limited to earth stations operating with non-Government space stations.

US320 Use of the 137–138, 148–149.9, and 400.15–401 MHz bands by the mobile-satellite service is limited to non-voice, non-geostationary satellite systems and may include satellite links between land earth stations at fixed locations.

US322 The 149.9–150.05 MHz band is allocated to the mobile-satellite service (Earth-to-space) on a primary basis after 1 January 1997 and shall be limited to non-voice, non-geostationary satellite systems, including satellite links between land earth stations. Before 1 January 1997 use of this band on a secondary basis for the mobile satellite service is allowed for land earth stations at fixed locations.

US323 In the 148–149.9 MHz band, no individual mobile earth station shall transmit, on the same frequency being actively used by fixed and mobile stations and shall transmit no more than 1% of the time during any 15 minute period; except, individual mobile earth stations in this band that do not avoid frequencies actively being used by the fixed and mobile services shall not exceed a power density of -16 dBW/4kHz and shall transmit no more than 0.25% of the time during any 15 minute period. Any single transmission from any individual mobile earth station operating in this band shall not exceed 450 ms in duration and consecutive transmissions from a single mobile Earth station on the same frequency shall be separated by at least 15 seconds. Land earth stations in this band shall be subject to electromagnetic compatibility analysis and coordination with terrestrial fixed and mobile stations.

US324 Government and non-Government satellite systems in the 400.15–401 MHz band shall be subject to electromagnetic compatibility analysis and coordination.

US325 In the band 148–149.9 MHz fixed and mobile stations shall not claim protection from land earth stations in the mobile-satellite service that have been previously coordinated; Government fixed and mobile stations exceeding 27 dBW EIRP, or an emission bandwidth greater than 38 kHz, will be coordinated with existing mobile-satellite service space stations.

US326 The 399.9–400.05 MHz band is allocated to the mobile-satellite service (Earth-to-space) on a primary basis after January 1, 1997 and shall be limited to non-voice, non-geostationary satellite systems, including satellite links between land Earth stations.

US327 The 2310–2360 MHz is allocated to the broadcasting-satellite service (sound) and complementary terrestrial broadcasting service on a primary basis. Such use is limited to digital audio broadcasting and is subject to the provisions of Resolution 528.

US328 In the band 2310–2360 MHz, the mobile and radiolocation services are allocated on a primary basis until 1 January 1997 or until a broadcasting-satellite (sound) service has been brought into use in such a manner as to affect or be affected by the mobile and radiolocation services in those service areas, whichever is later. The broadcasting-satellite (sound) service during implementation should also take cognizance of the expendable and reusable launch vehicle frequencies 2312.5, 2332.5, and 2352.5 MHz, to minimize the impact on this mobile service use to the extent possible.

US330 In the frequency bands 901–902 MHz, 930–931 MHz, and 940–941 MHz, the only fixed services permitted are ancillary services used in support of mobile personal communications services.

US331 In the frequency band 1850–1990 MHz, the only fixed PCS services permitted are ancillary services used in support of mobile personal communications services.

US334 In the band 17.8–20.2 GHz, Government space stations and associated earth stations in the fixed-satellite (space-to-Earth) service may be authorized on a primary basis. For a Government geostationary satellite network to operate on a primary basis, the space station shall be located outside the arc measured from East to West, 70° W to 120° W. Coordination between Government fixed-satellite systems and non-Government systems operating in accordance with the United States Table of Frequency Allocations is required.

USxxx In the band 13.75–13.8 GHz, earth stations in the fixed-satellite service shall be coordinated on a case-by-case basis with Government operations in order to minimize harmful interference to the Tracking and Data Relay Satellite System.

## GOVERNMENT FOOTNOTES

These footnotes, each consisting of the letter G followed by one or more digits, denote stipulations applicable only to the federal government. Only those government footnotes that pertain to allocations in 30 MHz–300 GHz are shown.

G2 In the bands 216–225, 420–450 (except as provided by US217) 890–902, 928–942, 1300–1400, 2300–2390, 2400–2402, 2417–2450, 2700–2900, 5650–5925 and 9000–9200 MHz, the Government radiolocation is limited to the military services.

G5 In the bands 162.0125–173.2, 173.4–174, 406.1–410 and 410–420 MHz, the fixed and mobile services are all allocated on a primary basis to the Government non-military agencies.

G6 Military tactical fixed and mobile operations may be conducted nationally on a secondary basis: (1) To the meteorological aids service in the band 403–406 MHz; and (2) to the radio astronomy service in the band 406.1–410 MHz. Such fixed and mobile operations are subject to local coordination to ensure that harmful interference will not be caused to the services to which the bands are allocated.

G8 Low power Government radio control operations are permitted in the band 420–450 MHz.

G11 Government fixed and mobile radio services, including low power radio control operations, are permitted in the band 902–928 MHz on a secondary basis.

G15 Use of the band 2700–2900 MHz by the military fixed and shipborne air defense radiolocation installations will be fully coordinated with the meteorological aids and aeronautical radionavigation services. The military air defense installations will be moved from the band 2700–2900 MHz at the earliest practicable date. Until such time as military air defense installations can be accommodated satisfactorily elsewhere in the spectrum, such operations will, insofar as practicable, be adjusted to meet the requirements of the aeronautical radionavigation service.

G19 Use of the band 9000–9200 MHz by military fixed and shipborne air defense radiolocation installations will be fully coordinated with the aeronautical radionavigation service, recognizing fully the safety aspects of the latter. Military air defense installations will be accommodated ultimately outside this band. Until such time as military defense installations can be accommodated satisfactorily elsewhere in the spectrum such operations will, insofar as practicable, be adjusted to meet the requirements of the aeronautical radionavigation services.

G27 In the bands 225–328.6, 335.4–399.9, and 1350–1400 MHz, the fixed and mobile services are limited to the military services.

G30 In the bands 138–144, 148–149.9, 150.05–150.8, 1427–1429, and 1429–1435 MHz, the fixed and mobile services are limited primarily to operations by the military services.

G31 In the bands 3300–3500 MHz, the Government radiolocation is limited to the military services, except as provided by footnote US108.

G32 Except for weather radars on meteorological satellites in the band 9975–10025 MHz and for Government survey operations (see footnote US108), Government radiolocation in the band 10000–10500 MHz is limited to the military services.

G34 In the band 34.4–34.5 GHz, weather radars on board meteorological satellites for cloud detection are authorized to operate on the basis of equality with military radiolocation devices. All other non–military radiolocation in the band 33.4–36.0 GHz shall be secondary to the military services.

G42 Space command, control, range and range rate systems for earth station transmission only (including installations on certain Navy ships) may be accommodated on a co–equal basis with the fixed and mobile services in the band 1761–1842 MHz. Specific frequencies required to be used at any location will be satisfied on a coordinated case–by–case basis.

G56 Government radiolocation in the bands 1215–1300, 2900–3100, 5350–5650 and 9300–9500 MHz is primarily for the military services; however, limited secondary use is permitted by other Government agencies in support of experimentation and research programs. In addition, limited secondary use is permitted for survey operations in the band 2900–3100 MHz.

G59 In the bands 902–928 MHz, 3100–3300 MHz, 3500–3700 MHz, 5250–5350 MHz, 8500–9000 MHz, 9200–9300 MHz, 13.4–14.0 GHz, 15.7–17.7 GHz and 24.05–24.25 GHz, all Government non–military radiolocation shall be secondary to military radiolocation, except in the subband 15.7–16.2 GHz airport surface detection equipment (ASDE) is permitted on a co–equal basis subject to coordination with the military departments.

G100 The bands 235–322 MHz and 335.4–399.9 MHz are also allocated on a primary basis to the mobile–satellite service, limited to military operations.

G101 In the band 2200–2290 MHz, space operations (space–to–Earth) and (space–to–space), and earth exploration–satellite (space–to–Earth) and (space–to–space) services, may be accommodated on a co–equal basis with fixed, mobile and space research services.

G104 In the bands 7450–7550 and 8175–8215 MHz, it is agreed that although the military space radio communication systems, which include earth stations near the proposed meteorological–satellite installations will precede the meteorological–satellite installations, engineering adjustments to either the military or the meteorological–satellite systems or both will be made as mutually required to assure compatible operations of the systems concerned.

G109 All assignments in the band 157.0375–157.1875 MHz are subject to adjustment to other frequencies in this band as long term U.S. maritime VHF planning develops, particularly that planning incident to support of the National VHF–FM Radiotelephone Safety and Distress System (See Doc. 15624/1-1.9.111/1.9.125).

G110 Government ground–based stations in the aeronautical radionavigation service may be authorized between 3500–3700 MHz where accommodation in the 2700–2900 MHz band is not technically and/or economically feasible.

G114 In the band 1350–1400 MHz, the frequency 1381.05 MHz with emissions limited to ± 12 MHz is also allocated to Fixed and Mobile-Satellite Services (space–to–Earth) for the relay of nuclear burst data.

G116 The band 7125–7155 MHz is also allocated for Earth–to–space transmissions in the Space Operations Service at a limited number of sites (not to exceed two), subject to established coordination procedures.

G117 In the bands 7.25–7.75 GHz, 7.9–8.4 GHz, 17.8–21.2 GHz, 30–31 GHz, 39.5–40.5 GHz, 43.5–45.5 GHz, and 50.4–51.4 GHz the Government fixed-satellite and mobile-satellite services are limited to military systems.

G118 Government fixed stations may be authorized in the band 1700–1710 MHz only if spectrum is not available in the band 1710–1850 MHz.

G120 Development of airborne primary radars in the band 2310–2390 MHz with peak transmitter power in excess of 250 watts for use in the United States is not permitted.

G122 The bands 2390–2400, 2402–2417 and 4660–4685 MHz were identified for immediate reallocation, effective August 10, 1994, for exclusive non-Government use under Title VI of the Omnibus Budget Reconciliation Act of 1993. Effective August 10, 1994, any Government operations in these bands are on a non-interference basis to authorized non-Government operations and shall not hinder the implementation of any non-Government operations.

G123 The bands 2300–2310 and 2400–2402 MHz were identified for reallocation, effective August 10, 1995, for exclusive non-Government use under Title VI of the Omnibus Reconciliation Act of 1993. Effective August 10, 1995, any Government operations in these bands are on a non-interference basis to authorized non-Government operations and shall not hinder the implementation of any non-Government operations.

G124 The band 2417–2450 MHz was identified for reallocation, effective August 10, 1995, for mixed Government and non-Government use under Title VI of the Omnibus Reconciliation Act of 1993.

# NON-GOVERNMENT FOOTNOTES

These footnotes, each consisting of the letters NG followed by one or more digits, denote stipulations applicable only to non-federal government stations. Only those non-government footnotes that pertain to allocations in 30 MHz–300 GHz are shown.

NG2 Facsimile broadcasting stations may be authorized in the band 88–108 MHz.

NG3 Control stations in the domestic public mobile radio service may be authorized frequencies in the band 72–73 and 75.4–76 MHz on the condition that harmful interference will not be caused to operational fixed stations.

NG4 The use of the frequencies in the band 152.84–153.38 MHz may be authorized, in any area, to remote pickup broadcast base and mobile stations on the condition that harmful interference will not be caused to stations operating in accordance with the Table of Frequency Allocations.

NG6 Stations in the public safety radio services authorized as of June 30, 1958, to use frequencies in the band 159.51–161.79 MHz in areas other than Puerto Rico and the Virgin Islands may continue such operation, including expansion of existing systems, on the condition that harmful interference will not be caused to stations in the services to which these bands are allocated. In Puerto Rico and the Virgin Islands this authority is limited to frequencies in the band 160.05–161.37 MHz. No new public radio service system will be authorized to operate on these frequencies.

NG12 Frequencies in the bands 454.40–455 MHz and 459.40–460 MHz may be assigned to domestic public land and mobile stations to provide a two-way air-ground public radiotelephone service.

NG17 Stations in the land transportation radio services authorized as of May 15, 1958 to operate on the frequency 161.61 MHz may, upon proper application, continue to be authorized for such operation, including expansion of existing systems, on the condition that harmful interference will not be caused to the operation of any authorized station in the maritime mobile service. No new land transportation radio service system will be authorized to operate on 161.61 MHz.

NG23 Frequencies in the band 2100–2200 MHz may also be assigned to stations in the international fixed public radio service located south of 25° 30' north latitude in the State of Florida and in U.S. Possessions in the Caribbean area, provided, however, no new assignments in the band 2150–2162 MHz will be made to such stations after February 25, 1974.

NG28 The frequency band 160.86–161.40 MHz is available for assignment to remote pickup base and remote pickup mobile stations in Puerto Rico and the Virgin Islands only on a shared basis with the land transportation radio service.

NG30 Stations in the international fixed public radiocommunication service in Florida, south of 25° 30' north latitude, may be authorized to use frequencies in the band 716–890 MHz on the condition that harmful interference will not be caused to the broadcasting service of any country. This is an interim allocation the termination of which will later be specified by the Commission when it is determined that equipments are generally available for use in bands allocated internationally to the fixed service.

NG31 Stations in the Rural Radio Service licensed for Basic Exchange Telecommunications Radio Service may be authorized to use some frequencies in the bands 816–820 MHz (fixed subscriber) and 861–865 MHz (central office or base), on a co-primary basis with private land mobile radio licensees, pursuant to Part 22 Subpart H.

NG41 Frequencies in the bands 3700–4200 MHz, 5925–6425 MHz, and 10.7–11.7 GHz may also be assigned to stations in the international fixed public and international control services located in U.S. Possessions in the Caribbean area.

NG42 Non-Government stations in the radiolocation service shall not cause harmful interference to the amateur service.

NG43 Fixed stations in the domestic public radio services in Alaska, south of 56° north latitude and east of 134° west longitude, may be authorized to use frequencies in the band 800–830 MHz, on the condition that harmful interference will not be caused to the broadcasting service of any country.

NG47 In Alaska, frequencies between the band 2655–2690 MHz are not available for assignment to terrestrial stations.

NG49 The following frequencies may be authorized for mobile operations in the Manufacturers Radio Service subject to the condition that no interference is caused to the reception of television stations operating on channels 4 and 5 and that their use is limited to a manufacturing facility:

MHz

72.02 72.06 72.10 72.14 72.18 72.22 72.26 72.30 72.34 72.38

72.04 72.08 72.12 72.16 72.20 72.24 72.28 72.32 72.36 72.40

Further, the following frequencies may be authorized for mobile operations in the Special Industrial Radio Service, Manufacturers Radio Service, Railroad Radio Service and Forest Products Radio Service subject to the condition that no interference is caused to the reception of television stations operating on channels 4 and 5; and that their use is limited to a railroad yard, manufacturing plant, logging site, mill, or similar industrial facility.

MHz

72.44 72.48 72.52 72.56 72.60 75.44 75.48 75.52 75.56 75.60

NG51 In Puerto Rico and the Virgin Islands only, the bands 150.8–150.98 MHz and 150.98–151.49 MHz are allocated exclusively to the business radio service.

NG53 In the band 12.7–13.15 GHz, television pickup stations and CARS pickup stations shall be assigned channels on a co-equal basis and shall operate on a secondary basis to fixed stations operating in accordance with the Table of Frequency Allocations. In the 13.15–13.20 GHz band television pickup stations and CARS pickup stations shall be assigned on an exclusive basis in the top one hundred markets, as set out in Section 76.51.

NG56 In the bands 72.0–73.0 and 75.4–76.0 MHz, the use of mobile radio remote control of models is on a secondary basis to all other fixed and mobile operations. Such operations are subject to the condition that interference will not be caused to common carrier domestic public stations, to remote control of industrial equipment operating in the 72–76 MHz band, or to the reception of television signal on channels 4 (66–72 MHz) or 5 (76–82 MHz). Television interference shall be considered to occur whenever reception of regularly used television signals is impaired or destroyed, regardless of the strength of the television signal or the distance to the television station.

NG59 The frequencies 37.60 and 37.85 MHz may be authorized only for use by base, mobile, and operational fixed stations participating in an interconnected or coordinated power service utility system.

NG63 Television Broadcast translator stations holding valid licenses on November 15, 1971, to operate in the frequency band 806–890 MHz (channels 70–83), may continue to operate in this band, pursuant to periodic license renewals, on a secondary basis to the land mobile radio service.

NG66 The frequency band 470–512 MHz is allocated for use in the broadcasting and land mobile radio services. In the land mobile services it is available for assignment in the domestic public, public safety, industrial, and land transportation radio services at, or in the vicinity of 13 urbanized areas of the United States, as set forth in the table below. Additionally, in the land mobile services, TV Channel 16 is available for assignment in the public safety radio services at, or in the vicinity of Los Angeles. Such use in the land mobile services is subject to the conditions set forth in Parts 22 and 90 of this chapter, CFR 47.

| Urbanized area | TV channel |
|---|---|
| New York-Northeastern New Jersey | 14, 15 |
| Los Angeles | 14, 20 |
| Chicago-Northwestern Indiana | 14, 15 |
| Philadelphia, PA-New Jersey | 19, 20 |
| Detroit, Michigan | 15, 16 |
| San Francisco-Oakland, California | 16, 17 |
| Boston, Mass | 14, 16 |
| Washington D.C.,-Maryland-Virginia | 17, 18 |
| Pittsburgh, PA | 14, 18 |
| Cleveland, Ohio | 14, 15 |
| Miami, Florida | 14 |
| Houston, Texas | 17 |
| Dallas, Texas | 16 |

NG70 In Puerto Rico and the Virgin Islands only, the bands 159.240–159.435 and 160.410–160.620 MHz are

also available for assignment to base stations and mobile stations in the special industrial radio service.

NG101 The use of the band 2500–2690 MHz by the broadcasting-satellite service is limited to domestic and regional systems for community reception of educational television programming and public service information. Such use is subject to agreement among administrations concerned and those having services operating in accordance with the table, which may be affected. Unless such agreement includes the use of higher values, the power flux density at the Earth's surface produced by emissions from a space station in this service shall not exceed those values set forth in Part 73 of the rules for this frequency band.

NG102 The frequency bands 2500–2655 MHz (space-to-Earth) and 2655–2690 MHz (Earth-to-space) are allocated for use in the fixed-satellite service as follows:

(a) For common carrier use in Alaska, for intra-Alaska service only, and in the mid and western Pacific areas including American Samoa, the Trust Territory of the Pacific Islands, Guam and Hawaii;

(b) For educational use in the contiguous United States, Alaska and the mid and western Pacific areas including American Samoa, the Trust Territory of the Pacific Islands, Guam and Hawaii.

Such use is subject to agreement with administrations having services operating in accordance with the table, which may be affected. In the band 2500–2655 MHz, unless such agreement includes the use of higher values, the power flux density at the Earth's surface produced by emissions from a space station in this service shall not exceed the values set forth in Part 25 of the Rules for this frequency band.

NG104 The use of the bands 10.7–11.7 and 12.75–13.25 GHz in the fixed satellite service is limited to international systems, i.e., other than domestic systems.

NG111 The band 157.4375–157.4625 MHz may be used for one way paging operations in the special emergency radio service.

NG112 The frequencies 25.04, 25.08, 150.980, 154.585, 158.445, 159.480, 454.000 and 459.000 MHz may be authorized to stations in the petroleum radio service for use primarily in oil spill containment and cleanup operations and secondarily in regular land mobile communication.

NG114 In the Gulf of Mexico offshore from the Louisiana-Texas coast, the frequency band 476–494 MHz (TV channels 15, 16 and 17) is allocated to the Domestic Public and Private Land Mobile Radio Services in accordance with the regulations set forth in Parts 22 and 90 respectively.

NG115 In the 174 to 216 MHz band wireless microphones may be authorized to operate on a secondary, non-interfering basis, subject to terms and conditions set forth in Part 74 of these Rules and Regulations.

NG117 The frequency 156.050 and 156.175 MHz may be assigned to stations in the maritime mobile service for commercial and port operations in the New Orleans Vessel Traffic Service (VTS) area and the frequency 156.250 MHz may be assigned to stations in the maritime mobile service for port operating in the New Orleans and Houston VTS areas.

NG118 Television translator relay stations may be authorized to use frequencies in this band on a secondary basis to stations operating in accordance with the Table of Frequency Allocations.

NG120 Frequencies in the 928–960 MHz band may be assigned for multiple address systems and mobile operations on a primary basis as specified in Part 94 of this chapter.

NG122 Television Pickup stations may be authorized under Part 74 in the 6425–6525 MHz band on a secondary basis to stations operating in accordance with the Table of Frequency Allocations.

NG124 In the public safety radio service allocations within the bands 30–50 MHz, 150–174 MHz and 450–470 MHz, police radio service licenses are authorized to operate low powered radio transmitters on a secondary, non-interference basis in accordance with the provisions of Sections 2.803 and 90.19(f)(5) of the Rules.

NG127 In Hawaii, the frequency band 488–494 MHz is allocated exclusively to the fixed service for use by common carrier control and repeater stations for point-to-point inter-island communications only.

NG128 In the 535–1705 kHz band, AM broadcast licensees or permittees may use their AM carrier on a secondary basis to transmit signals intended for both broadcast and non-broadcast purposes. In the 88–108 MHz band, FM broadcast licensees or permittees are permitted to use subcarriers on a secondary basis to transmit signals intended for both broadcast and nonbroadcast purposes. In the 54–72, 76–88, 174–216 and 740–890 MHz bands, TV broadcast licensees or permittees are permitted to use subcarriers on a secondary basis for both broadcast and non-broadcast purposes.

NG129 In Alaska, the bands 76–88 MHz and 88–100 MHz are also allocated to the Fixed service on a secondary basis. Broadcast stations operating in these bands shall not cause interference to non-Government fixed operations authorized prior to January 1, 1982.

NG134 In the band 10.45–10.5 GHz non-Government stations in the radiolocation service shall not cause harmful interference to the amateur and amateur-satellite services.

NG135 In the 420–430 MHz band the amateur service is not allocated north of line A (def. Sec. 2.1).

NG139 Pending adopting of further specific rules concerning usage of the band 12.2–12.7 GHz by the fixed and broadcasting-satellite services, systems in these services may be authorized subject to the condition that adjustments in certain systems design or technical parameters may become necessary during the systems lifetime. The necessity for such adjustments, and their extent, will be dependent upon the Final Acts of the 1983 Regional Administrative Radio Conference and subsequent Commission decisions.

NG140 Pending adopting of further specific rules concerning usage of the band 17.3–17.8 GHz by the fixed-satellite service for the purpose of providing feeder links to the broadcasting-satellite service, systems may be authorized for this purpose subject to the condition that adjustments in certain systems design or technical parameters may become necessary during the system lifetime. The necessity for such adjustments, and their extent, will be dependent upon the Final Acts of the 1983 Regional Administrative Radio Conference and subsequent Commission decisions.

NG141 The frequencies 42.40 MHz and 44.10 MHz are authorized on a primary basis in the State of Alaska for meteor burst communications by fixed stations in the Rural Radio Service operating under the provisions of Part 22 of this Chapter. The frequencies 44.20 MHz and 45.90 MHz are authorized on a primary basis in Alaska for meteor burst communications by fixed private radio stations operating under the provisions of Part 90 of the Chapter. The private radio station frequencies may be used by Common Carrier stations on a secondary, non-interference basis and the Common Carrier frequencies may be used by private radio stations for meteor burst communications on a secondary, non-interference basis. Users shall cooperate to the extent practical to minimize potential interference. Stations utilizing meteor burst communications shall not cause harmful interference to stations of other radio services operating in accordance with the Table of Frequency Allocations.

NG143 In the band 11.7–12.2 GHz protection from harmful interference shall be afforded to transmissions from space stations not in conformance with international footnote 839 only if the operations of such space stations impose no unacceptable constraints on operations or orbit locations of space stations in conformance with 839.

NG144 Stations authorized as of September 9, 1983 to use frequencies in the band 17.7–19.7 GHz may, upon proper application, continue to be authorized for such operation.

NG145 In the band 11.7–12.2 GHz, transponders on space stations in the fixed-satellite service may be used additionally for transmissions in the broadcasting-satellite service, provided that such transmissions do not have a maximum e.i.r.p. greater than 53 dBW per television channel and do not cause greater interference or require more protection from interference than the coordinated fixed-satellite service frequency assignments. With respect to the space services, this band shall be used principally for the fixed satellite service.

NG147 Stations in the broadcast auxiliary service and private radio services licensed as of July 25, 1985, or on a subsequent date following as a result of submitting an application for license on or before July 25, 1985, may continue to operate on a radiodetermination satellite service.

NG148 The frequencies 154.585 MHz, 159.480 MHz, 160.725 MHz, 160.785 MHz, 454.000 MHz and 459.000 MHz may be authorized to maritime mobile stations for offshore radiolocation and associated telecommand operations.

NG149 The frequency bands 54–72 MHz, 76–88 MHz, 174–216 MHz, 470–512 MHz, 512–608 MHz, and 614–806 MHz are also allocated to the Fixed Service to permit subscription television operations in accordance with Part 73 of the rules.

NG151 In the frequency bands 824–849 MHz and 869–894 MHz, cellular land mobile licensees are permitted to offer auxiliary services on a secondary basis subject to the provisions of Part 22.

NG152 The band 219–220 MHz is also allocated to the amateur service on a secondary basis for stations participating, as forwarding stations, in point-to-point fixed digital message forwarding systems, including intercity packet backbone networks.

NG153 The 2110–2150 MHz and 2160–2200 MHz bands are reserved for future emerging technologies on a co-primary basis with the fixed and mobile services. Allocations to specific services will be made in future proceedings.

# Subject Index

## A

## B



## C

## D



## E

## F



## G

## M



## N

## Q



## R

## U



## V

## W



## Z

# BAND INDEX

## About the Author

Bennett Kobb is a writer and researcher for electronics manufacturers, FCC licensees and publishers.

He is a founder of the Wireless Information Networks Forum, an association for developers of unlicensed Personal Communications Service devices.

He founded *Federal Communications TechNews* (now *Telecommunications Reports: Wireless News*) and was editor of *Cellular Radio News* and *Personal Communications* magazine, the first periodicals for the cellular telephone industry. His articles and reports appear in numerous trade and consumer newsletters and magazines.

He is a Fellow of the Radio Club of America and a member of the Institute of Electrical and Electronics Engineers. His Amateur Radio callsign is KC5CW.

Bennett received his M.S. in telecommunications from the University of Colorado and B.S. in radio-television-film from the University of Texas.

When he is not researching the airwaves, he may be found amid flying objects at the U.S. Department of Juggling, a Washington, D.C. society for the circus arts.

## *How to Order SpectrumGuide*

SpectrumGuide is available from technical booksellers and from Publication Services, Inc. (PSI), an authorized direct sales agent for New Signals Press.

To order from PSI with a credit card, phone:
800 460 0090 or 512 795 5006

Fax:
512 478 9263

Mail:
Publication Services, Inc.
824 W. 10th St. Suite 100
Austin TX 78701

Enclose credit card information or your check or money order for $30.00 for each book plus shipping and handling.

Book Rate shipping and handling to U.S. addresses:
$4.00 for the first book and $1.50 for each additional book.

Priority Mail shipping and handling (recommended) to U.S. addresses:
$6.00 for the first book and $3.50 for each additional book.

Orders to Virginia addresses must include 4.5% sales tax.

Visa, MasterCard, Discover, and American Express cards accepted. Please make checks and money orders payable to Publication Services, Inc. Checks must be in U.S. funds drawn on a U.S. bank.

Please contact PSI about shipping options, order status, resale, quantity sales, and other SpectrumGuide products.